Toward a History of the
Space Shuttle

An Annotated Bibliography

Compiled by
Roger D. Launius and Aaron K. Gillette

NASA History Office
Code JCH
NASA Headquarters
Washington, DC 20546

MONOGRAPHS IN AEROSPACE HISTORY
Number 1
December 1992

PREFACE

Since the idea of a reusable rocket-plane was first seriously-studied by Eugen Sänger in the 1930s, the concept has exerted strong influence on the development of human spaceflight. In the United States, detailed proposals for a reusable space vehicle were developed as early as the 1950s, and several projects reached the design and test stage in the 1960s. Initially, the Space Shuttle was envisioned as a fully reusable, commercial spaceplane. During the early 1970s, however, its development faced considerable obstacles, budgetary shortfalls, some congressional opposition, increasing public apathy, and design difficulties. What emerged was a smaller, semi-reusable vehicle, advertised as an economical and efficient means of space transport. Whether the Shuttle has fulfilled these goals is a topic of some controversy. Even so, the Space Shuttle has been the cornerstone of the U.S. space program, and the driving force behind much of the budget and programs of NASA for over two decades.

Throughout the long history of the Space Shuttle concept, numerous books, studies, reports, and articles have been written. This selective, annotated bibliography discusses those works judged to be most essential for researchers writing scholarly studies on the Space Shuttle's history. A thematic arrangement of material concerning the Shuttle will, it is hoped, bring clarity and simplicity to such a complex subject. Subjects include the precursors of the Shuttle, its design and development, testing and evaluation, and operations. Other topics revolve around the *Challenger* accident and its aftermath, promotion of the Shuttle, science on the Shuttle, commercial uses, the Shuttle's military implications, its astronaut crew, the Shuttle and international relations, the management of the Shuttle program, and juvenile literature. Along with a summary of the contents of each item, judgments have been made on the quality, originality, or importance of some of these publications. An index concludes this work.

The authors would like to acknowledge the assistance of those individuals who aided in the preparation of this bibliography. Lee D. Saegesser was instrumental in obtaining those documents listed below; J.D. Hunley edited and critiqued the text; Patricia Shephard typed the manuscript; the staffs of the NASA Headquarters Library and the Scientific and Technical Information Program provided assistance in locating bibliographical materials; and the NASA Headquarters Printing and Graphics Office developed the layout and handled printing.

This is the first publication in a new series of special studies prepared by the NASA History Office. The MONOGRAPHS IN AEROSPACE HISTORY series is designed to provide a wide variety of studies relative to the history of aeronautics and space. This series' publications are intended to be tightly focused in terms of subject, relatively short in length, and reproduced in an inexpensive format to allow timely and broad dessimination to researchers in aerospace history. Suggestions for additional publications in the MONOGRAPHS IN AEROSPACE HISTORY series are welcome.

ROGER D. LAUNIUS
Chief Historian
National Aeronautics and Space Administration

TOWARD A HISTORY OF THE SPACE SHUTTLE

CHAPTER 1
GENERAL WORKS

Baker, David. "A Schedule for the Shuttle." *Spaceflight*. 13 (December 1971) : 454 -55. Describes the timetable for completion and flight of the Shuttle as it was understood in the early 1970s. It comments that the Shuttle should be flying operational missions by 1979, but Baker suggests that there was already a contingency plan being developed to push flight operations as far into the future as the 1981-1983 time period depending on NASA's funding.

Baker, David. "Evolution of the Space Shuttle, Part 1." *Spaceflight*. 15 (June 1973): 202-10. First in a series of articles dealing with the development of the Shuttle. It focuses on the technological development of the system through the proposed baseline for Phase B studies.

Baker, David. "Evolution of the Space Shuttle, North American Rockwell, Part 2." *Spaceflight*. 15 (July 1973): 264-68. Second in series of articles, this one reviews the studies in the late 1960s on the possibility of a fully-reusable Shuttle, emphasizing technological development.

Baker, David. "Evolution of the Space Shuttle, North American Rockwell, Part 3." *Spaceflight*. 15 (September 1973): 344-52. The third Shuttle article in this series deals with the technological development of the external fuel tanks for the orbiter and the replacement of the piloted flyback booster with a ballistic booster arrangement.

Baker, David. " Evolution of the Space Shuttle, Part 4 ." *Spaceflight*. 18 (September 1976) : 304 -38. Written to commemorate the roll-out of the Orbiter 101 prototype, this article reviews the major design changes and presents an evolutionary update of the Shuttle's technological development.

Baker, David. "Evolution of the Space Shuttle: 5. Approach and Landing Test Programme." *Spaceflight*. 19 (June 1977): 213-17. As the subtitle suggests, this article describes the major developments in the test program for the Shuttle's reentry and landing phase.

Baker, David. "Evolution of the Space Shuttle: 6. Free Flight Tests Begin." *Spaceflight*. 20 (January 1978): 21-28, 40. Describes the flight characteristics of the Shuttle tests then underway at the Dryden Flight Research Center in California.

Baker, David. "Evolution of the Space Shuttle: 7. External Tank Design-1." *Spaceflight*. 20 (February 1978): 60-66. An article relating the basic development of the technology going into the external tank design.

Baker, David. "Evolution of the Space Shuttle: 8. External Tank Design-2." *Spaceflight*. 20 (March 1978): 110-15. A continuation of the above article.

Baker, David. *Space Shuttle*. New York: Crown Publishers, 1979. A popularly-written work on the origins and development of the Shuttle. Heavily illustrated, it bears a striking resemblance to the series of articles on the subject listed above.

Bekey, I., and Mayer, H. "1980-2000: Raising Our Sights for Advanced Space Systems." *Astronautics and Aeronautics*. 14 (July-August 1976): 34-64. This special section on the possible future of space travel has much to say about space stations and flights to Mars and beyond, but also emphasizes the development of the Shuttle as the necessary first step in any ready access to space.

Champine, Gloria R. *Langley's Space Shuttle Technology—A Bibliography*. Hampton, VA: Langley Research Center, 1981. A compilation of most of the major research reports, journal articles, presentations, and contractor reports written or published by the Langley Research Center staff or by its contractors. It covers a number of disciplines: aer othermodynamics, s tructures, d ynamics an d aer oelasticity, environment, a nd m aterials. Organized ch ronologically w ithin t hree m ajor cat egories—NASA formal reports, c ontractor r eports, a nd articles an d co nferences—the bibliography collects the material p roduced o nly th rough th e a uspices o f Langley. It is not annotated and there is no indication of any central location where all the listed items might be found. There are more than one thousand entries in this bibliography, most of them unpublished in the strict sense of the term—reports and technical papers which might have been duplicated but not generally made available—and all of them highly technical.

Collins, Michael. *Liftoff: The Story of America's Adventure in Space.* New York: Grove Press, 1988. This book, a general history of the U.S. space program for a popular audience written by a former astronaut, has a fine discussion of the development and flight of the Space Shuttle. He sketches, in an easy to understand style, the design and engineering development of the system, difficulties overcome, and the operations of the Shuttle. The m ost v aluable p art is C ollins' a nalysis o f th e *Challenger* accident. It w as p artly a p ersonal acco unt, describing his predictions and reactions, and presenting a clear portrait of the technical problem that caused the disaster. He then comments on NASA as an organization and offers some insights on how to bounce back from the tragedy.

Cooper, Henry S.F. "A R eporter A t L arge: S huttle-I." *The N ew Y orker.* 9 F ebruary 1981, pp. 43 -105, *passim.* A sophisticated account of the development of the Shuttle, written with attention to detail, human involvement, and real style. It surveys the course taken by the space program that led to the building of the Shuttle. Cooper makes f requent co mparisons b etween t he S huttle an d ear ly s pacecraft lik e A pollo, n ot j ust th e s earch f or reusable systems ala Sänger.

Cooper, Henry S.F. "A Reporter At Large: Shuttle-II." *The New Yorker.* 16 February 1981, pp. 65-113, *passim.* Second in the C ooper s eries, f ollows o n w ith c ontinued d iscussion o f the Shuttle an d s ome s peculation o n w hat the Shuttle might mean for the world and future space programs.

Covault, Craig. "Columbia R eady f or F irst F light." *Aviation W eek & Spac e T echnology.* 6 A pril 1981, pp. 16 -20. A description of the Shuttle, how it was developed, how it operates, and what promise it holds. This article was published at the time of the first launch.

Disher, John H. "Space Transportation: Reflections and Projections." in Durant, Frederick C., III, ed. *Between Sputnik and the Shuttle: New Perspectives on American Astronautics.* San Diego, CA: American Astronautical Society, 1981. pp. 199-224. This article is part of a larger publication focusing on various aspects of the space program. A presentation by the director of Advanced Programs for NASA's Office of Space Transportation Systems at the AAS, it contains no notes or other scholarly apparatus. It does survey the methods of spaceflight for piloted missions since Mercury and describes some of the features of the Shuttle.

Dooling, Dave. "Shuttle Business and S pace P olicy," *Astronautics & Aeronautics.* 19 (September 1981): 9-15. This article is a good, short assessment of the Shuttle mission and how it affects the commercial and military worlds; it raises some policy questions concerning the NASA's role in the operational arena.

Embury, Barbara, with Crouch, Tom D. *The Dream is Alive.* New York: Harper and Row, 1991. This book is a printed version of the National Air and Space Museum's IMAX film by the same title. Using powerful illustrations and spare text, it deals with the potential of the Shuttle.

Fletcher, James C. "Are SKYLAB and the Space Shuttle Worth the Investment?" *Government Executive.* January 1974. pp. 38 -40, 4 2. T his is a lo gical d efense o f b oth th e S huttle a nd S kylab p rograms w ritten b y the N ASA Administrator. H e j ustifies the S huttle on th e basis o f cost savings ($1 billion p er year for operations) and versatility, and its ability to serve as an excellent platform for scientific research, to mate humans and machines in a reliable and meaningful way, to mix scientific and practical applications, to provide a space rescue capability, to be used for Department of Defense projects, to foster aerospace technology, to facilitate international cooperation, and to provide the key to U.S. supremacy in space.

Forres, George. *Space Shuttle: The Quest Continues*. London: Ian Allen, 1989. In what could only be considered a broad introduction to the Shuttle program, the author describes the reassessment of the Shuttle program following the *Challenger* accident and its return to flight in September 1988. Designed for the buff market, it is well-illustrated but has no references.

Furniss, Tim. *Space Shuttle Log*. New York: Jane's, 1986. This is a decent general history of the Shuttle, promotional and heavily-illustrated in a large format.

Gatland, Kenneth W.; Hewish, Mark; and Wright, Pearce. *The Space Shuttle Handbook*. New York: Hamlyn, 1979. This is a short, heavily-illustrated work about the development and operation of the Shuttle, designed for the popular market.

Hawkes, Nigel. *Space Shuttle*. New York: Gloucester Press, 1982. A short work using drawings and diagrams, this booklet describes the new era of space technology that began officially in 1981 with the maiden flight of the Shuttle *Columbia*, the first reusable spacecraft.

Hawkes, Nigel. *Space Shuttle: A New Era?* New York: Gloucester Press, 1989. Rehashes Hawkes' 1982 publication but also discusses the evolution of the Shuttle program since first flight, including the *Challenger* accident. In this book the author, who had been a booster of the Shuttle program, takes a much more critical approach toward the effort and suggests that the Shuttle does not really provide easy access to space.

Henry, Beverly Z., and Decker, John P. "Future Earth Orbit Transportation Systems/Technology Implications." *Astronautics & Aeronautics*. 14 (September 1976): 18-28. Sophisticated article on the potential of the Shuttle for space exploration.

Hosenball, S. Neil. "The Space Shuttle: Prologue or Postscript?" *Journal of Space Law*. 9 (Spring-Fall 1981): 69-75. This article treats the development of the Shuttle as a method for easy access to space, focusing on the problems and potential of space commercialization, the legal issues of orbiting civilians, and associated questions. As might be expected, it is heavy on policy and legal questions and short on technological discussions.

Jenkins, Dennis R. *Rockwell International Space Shuttle*. Arlington, TX: Aerofax Inc., 1989. One of several popular books that present an overview of the Shuttle and its development and use. It is a well-illustrated, 72 page book.

Joels, Kerry Mark, and Kennedy, Greg. *Space Shuttle Operator's Manual*. New York: Ballantine Books, 1982. One of the better promotional-oriented, illustrated histories, this book describes the origins and development of the Shuttle.

Johnson, Thomas H. "The Natural History of the Space Shuttle." *Technology and Society*. 10 (1988): 417-24. Describes the historical development of the Shuttle from its inception to the present day. Johnson argues that "scientific rationality," as evidenced in the design of the system, was overridden by one president who ignored professional advisors, and by a second president who overruled DOD's protest of the Shuttle's monopoly on launches. He notes that mixing explicit technical and economic goals with subjective criteria of politics and prestige established a program in which shortcuts were common and the potential for failure downroad was probable. He notes that stress on competition promoted what he calls "irrational elements in decision making." In other words, Johnson maintains that cautious judgment and detailed analysis took a back seat to glamour and national prestige in the development of the Shuttle.

"Just How Does the Shuttle Stack Up?" *NASA Activities*. 21 (November/December 1990): 8-10. Attempts to justify the Shuttle by comparing its performance in the 1980s with previous human space flight missions.

Kaplan, Marshall H. *Space Shuttle: America's Wings to the Future*. Fallbrook, CA: Aero Publishing, 1978. A popular treatment, with heavy reliance on photographs and a public relations tone, that discusses the development of the Shuttle from a nontechnical perspective. The subtitle summarizes the thrust of the work; useful chiefly as an example of the public image of the space program.

Larmore, Lewis, and Gervais, R.L. eds. *Space Shuttles and Interplanetary Missions*. Tarzana, CA: American Astronautical Society, 1970. This lengthy publication is a collection of papers presented at an AAS meeting on the use of the Shuttle, then only on the drawing board, for flights to Mars and other planets. These discussions were highly speculative.

Light, Larry. "Space Shuttle Will Usher in New Spaceflight Era...Despite Delays, Budget Cuts and Controversy." *Congressional Quarterly*. 37 (28 April 1979): 784-5. Reviews the congressional battle over funding for the Shuttle, and summarizes the Shuttle's expected military, scientific, and commercial uses.

Luxenberg, Barbara A. "Space Shuttle Issue Brief #IB73091." The Library of Congress Congressional Research Service Major Issues System. 7 July 1981. Describes the general development and mission of the Shuttle and provides commentary on issues of importance in the political arena of the period, especially relating to operational performance and costs.

Lyndon B. Johnson Space Center. *Space Shuttle*. Washington DC: National Aeronautics and Space Administration SP-407, 1976. This highly-illustrated, full-color booklet was written within NASA as a means of convincing the public of the advantages of the Shuttle program. It describes the Shuttle system and mission profile, the technology required, the benefits accrued from living and working in space, and the major components of the Shuttle. The last sections deal with the economic impact of the Shuttle and the field center participants. Although now 16 years old it is still a useful summary of the Shuttle and its capabilities and potential uses.

MacKnight, Nigel. *Shuttle*. Osceola, WI: Motorbooks International, 1985. A heavily illustrated, popular book on the Shuttle.

MacKnight, Nigel. *Shuttle 2*. Osceola, WI: Motorbooks International, 1988. Revision of the 1985 edition, with more illustrations and a commentary on the *Challenger* accident.

Mark, Hans. "The Impact of Our Enterprise in Space." *Technology in Society*. 1 (1979): 43-53. Broad-brush assessment of the nation's space program. Mark, a former top NASA official, pays special attention to the Shuttle program as the best effort the U.S. has for routine access to space. Once ready access is achieved, the door will be open for grand developments in science and the applications of science and technology in new areas, it argues.

National Aeronautics and Space Administration. *National Space Transportation System Reference*. Washington, DC: National Aeronautics and Space Administration, June 1988. 2 Vols. Contains a wealth of information about the Shuttle. The first volume has material about systems and facilities and the second is about Shuttle operations.

National Aeronautics and Space Administration. *National Space Transportation System Overview*. Washington, DC: National Aeronautics and Space Administration, September 1988. Brief description of the overall Shuttle program, assessing the status of each major aspect of the effort and offering a soothing statement that problems were corrected after the *Challenger* accident and by 1988 the Shuttle program was running well.

National Aeronautics and Space Administration. *Space Shuttle News Reference*. Washington, DC: National Aeronautics and Space Administration, 1981. This is a loose leaf reference source about the Shuttle designed for use by media. It has basic facts about all manner of subjects and approaches the topic with a characteristic public relations slant common to such types of material.

National Commission on Space. *Pioneering the Space Frontier: The Report of the National Commission on Space*. New York: Bantam Books, 1986. Plays on the theme of the pioneering spirit of America and sets forth the proposition that the nation will go into space, thus opening the next frontier. The book asks questions about how the U.S. will meet the challenge of space exploration in the next century, in the process developing an aggressive program for a space station and colonies on the Moon and eventually Mars. The centerpiece of this program, however, is the ability to reach space economically and routinely. In that environment a mixture of expendable launch vehicles and the Shuttle are recommended. The commission urges the development of more efficient space transportation systems through technological research, especially fully reusable vehicles of all types. A chapter, "Highway to Space," deals specifically with the Shuttle and urges its continued development. It also recommends the development of a follow-on system, an aerospace plane that could take off and land like an aircraft but reach orbit.

O'Neill, Gerard K. *The High Frontier: Human Colonies in Space*. New York: William Morrow and Co., 1977. Speculates on what the universe and humanity will be like in the 21st century. A central part of the book is movement into space made possible by the Space Shuttle.

Oberg, James E. "Beyond the Space Shuttle." *Astronomy*. 4 (March 1976): 6-19. This article speculates on the potential for space exploration opened by the Shuttle. It discusses the possibilities of a space station and trips to the other planets.

Parkinson, R.C. "Earth-Moon Transport Options in the Shuttle and Advanced Shuttle Era." *Journal of British Interplanetary Society*. 34 (February 1981). This article is a useful discussion of the potential operational activities of the Shuttle in an environment in which access to space was assured, in which there was a space station, and in which there was a Moon base.

Pielke Jr., Roger A. "A Reappraisal of the Space Shuttle Program." Unpublished study conducted by the Center for Space and Geosciences Policy, University of Colorado. Copy in NASA History Office Reference Collection. Part 1 critiques the "conventional wisdom" concerning the Shuttle's development, that the Space Shuttle is a "policy failure" because NASA bowed to political pressure in designing a less costly but also less effective shuttle. It criticizes the "Apollo paradigm" that a clear goal and strong presidential commitment to a large space program is necessary for success. Rather, the article argues, the space program should adapt itself to the system of incremental politics that characterizes U.S. government. Using the level of federal commitment and program performance with respect to original expectations as the main criteria for success, Pielke concludes that the Shuttle program has performed poorly. Part II of Pielke's study attempts to identify the specific causes of the Shuttle's "poor performance." Pielke claims that the problem stems from NASA's control over technical expertise in astronautics, NASA's fixation on the Shuttle as a stepping-stone to larger programs, changing justifications for the Shuttle in order to please Congress and the Office of Management and Budget, pork-barrel politics, and NASA's expectation of non-accountability.

Powers, Robert M. *Shuttle: The World's First Spaceship*. Harrisburg, PA: Stackpole Books, 1979, 1980. A popular work written in the excited tones of the early Shuttle period, this book lays out the mission and direction of the program as it stood at the end of the 1970s. There is quite a lot of ballyhoo about what the Shuttle should be able to accomplish, "routine access to space," and what that would mean for the world. There is a short bibliography at the end of the book.

Priestley, Lee. *America's Space Shuttle*. New York: J. Messner, 1978. Written for a popular audience and containing numerous illu strations, th is b ook d iscusses th e r eusable s pace tr ansportation s ystem from a routine a nd pedestrian perspective.

Rector, W illiam F., III, and Penzo, Paul A., eds. *Space Shuttle: Dawn of An Era*. S an Diego, C A: U nivelt Inc., 1 980. This is a collection of essays on the Shuttle's promise for routine access to space. It contains the proceedings of the 26th annual meeting of the American Astronautical Society.

Roland, Alex. " The S huttle: T riumph o r T urkey?" *Discover*. November 1985, pp. 14 -24. W ritten b y a c ritic o f th e Shuttle program, Roland argues that the Shuttle was sold as a p ractical and cost-effective way to gain routine access to space. It has not delivered. It is still in the spectacle stage and its much-touted capabilities have not been realized. It has made far fewer flights and conducted fewer scientific experiments than publicly predicted.

Salkeld, R obert, P atterson, D onald W ., an d G rey, J erry. *Space Transportation Sy stems: 1980 -2000*. Ne w Yo rk: American Institute of Aeronautics and Astronautics, 1978. This s hort b ook d eals with the c oncept, development, building, and proposed use of the Space Shuttle. It has a section dealing with the origins of the reusable spacecraft concept and an assessment o f the u ses of the Shuttle. It also recommends t he vigorous pursuit of the goals of the Shuttle, largely because of the practical and economic benefits that will accrue.

Sharpless, J ack. *The Earthbound Observer: A P ersonal Look at t he P eople of N ASA and the Spac e Shut tle Effort*. Michigan City, IN: Sharpless Corp., 1983. This is an eclectic critique of the Shuttle program in 89 pages.

Smith, E.P. "Space Shuttle in Perspective—History in the Making." *AIAA P aper 75-336*. February 24-26, 1975, pp. 1- 13. This survey reviews four decades of space vehicle s ystems r esearch an d d evelopment l eading u p t o the building o f th e S huttle. I t tr aces th e h ighlights o f th ose d evelopments le ading to a reusable orbiter from Wernher von Braun's winged A-4b and Eugen Sänger's antipodal bomber of the World War II era, through the X-series o f t est ai rcraft an d r ocket p lanes i n the U.S. S mith p ays s pecial at tention t o the d esign s tudies and hardware programs of the 1950s and 1960s from which the Shuttle emerged. Finally, he recites the major steps in Shuttle review from its initial planning stage to the RDT&E process of the 1970s. This is a very informative, scholarly article.

Smith, Melvyn. *An Illustrated History of the Space Shuttle*. Sommerset, England: Haynes Pub. Group, 1985. This large- sized picture book is oriented toward satisfying the popular market. Almost half of it is concerned with earlier high-speed, hi gh-altitude f light a s a m eans o f p aving th e w ay f or th e S huttle. I t r ecites a nd publishes photographs of early aircraft, such as the X-1, the X-15, and lifting body studies before going into a discussion of t he S huttle. T his d iscussion focuses o n the t echnological development o f the o rbiter, es pecially test an d evaluation. A chapter is th en d evoted to e ach o f th e S huttle o rbiters b uilt, d ealing w ith th eir p rocurement, construction, test and evaluation, and mission performance. There is a helpful set of appendices discussing each of the X-15, M2F2, HL-10, X-24, M2F3, and Shuttle flights. There are no references.

Soffen, Gerald A. ed. *Visions of Tomorrow: A Focus on National Space Transportation Issues*. San Diego, CA: Univelt, 1987. This contains the proceedings of the 25th Goddard Memorial Symposium held 18-20 March 1987 at the Goddard Space Flight C enter, G reenbelt, Mar yland, an d sponsored b y t he American Astronautical Society. There are several articles on what the future m ight h old for s pace transportation as well as practical areas of exploitation for the Shuttle.

Stine, G. Harry. *Shuttle into Space: A Ride in America's Space Transportation System*. Chicago: Follett Publishing Co., 1978. This is an interesting account of the development and especially the promise offered by the Shuttle.

Stine, G. Harry. *The Hopeful Future*. New York: The Macmillan Co., 1983. In this book one of the leading "futurists" in the nation analyzes what humanity can accomplish in the twenty-first century. His emphasis is on science and technology, and he discusses at length the need to move outward into the solar system. He contends that many of the world's present problems can be solved by the exploration, and in some instances exploitation, of space. Much of the species' reach for the planets hinges on the Shuttle and its ability to support a reasonably-priced and routine entry into space.

Thompson, R.F. *The Space Shuttle: A Future Space Transportation System*. Houston, TX: Johnson Space Center, 1974. Originally presented as an AIAA paper, this short work sets forth the primary objective of the Space Shuttle program as an attempt to achieve an economical means to reach space. It provides an introductory review of the considerations that led to NASA's development of the Shuttle; describes the historical context for this discussion from the standpoint of general developments in transportation; and presents a review of the Shuttle system, mission profile, payload categories, and payload accommodations. It concludes with a forecast of the system's use for space science research.

Torres, George. *Space Shuttle: A Quantum Leap*. Novato, CA: Presidio Press, 1987. Discusses the Shuttle and the technology that it brings to space flight. Well-illustrated and packaged in a format designed for a popular audience.

Torres, George. *Space Shuttle: The Quest Continues*. Novato, CA: Presidio Press, 1989. A revision of Torres' 1987 publication on the Shuttle. Adds some additional detail and includes a section on the *Challenger* accident.

Turner, Sarah. "Maxime Faget and the Space Shuttle." *NASA Activities*. 21 (November/December 1990): 22. This short biography of Faget credits him with making major contributions toward developing the concepts and designs of every U.S. manned spacecraft from the Mercury program to the Space Shuttle.

"A Walk Around the Space Shuttle." *NASA Activities*. 21 (November/December 1990): 16-17. This short article provides a thumb-nail sketch of the main components of the Space Shuttle: the orbiter, engines, external tank, and solid rocket boosters.

Wilford, John Nobel. "Riding High." *Wilson Quarterly*. 4 (Autumn 1980): 56-70. Lead article in a special section of the space program. It details the recent history of the United States space program since the launch of Sputnik and the declaration of the space race by Kennedy. The author examines the interrelationships involved in funding, planning, and administration of the space program between Congress, NASA, and the DOD. He also examines highlights of the Mercury, Gemini, and Apollo projects, as well as the development of administrative techniques for management. Wilford suggests that the 1970s was a decade of decline for interest in the space program but that with the development of the Shuttle this appears to be changing. The present effort, he says, seems to represent an important combination of military, commercial, and scientific interests that have not been present in the space program before.

Wilson, Andrew. *Space Shuttle Story*. New York: Hamlyn, 1986. Large-format, illustrated history of the Shuttle designed mostly for the popular market.

Yenne, Bill. *Space Shuttle*. New York: Gallery Books, 1989. Another illustrated history with excellent color photos and simple text.

Link to Part 2 (1992–2011), Chapter 1—General Works

CHAPTER 2
PRECURSORS OF THE SHUTTLE

Akridge, Max. "Space Shuttle." 8 January 1970. Document in the NASA History Office Reference Collection. Written by a member of the Marshall Space Flight Center's Space Shuttle Task Team, this monograph-length work, without footnotes, discusses the history of the Space Shuttle concept from 1950 through the Integral Launch and Reentry Vehicle Studies (Phase A). It addresses the economics, marketability, conceptualization, and preliminary design of reusable launch vehicles as devised in various "Space Shuttle" studies.

"Ancestors of the Space Shuttle: The X-15, Dyna-Soar, X-23 and X-24 Programs." *Space World: The Magazine of Space News*. December 1978, pp. 4-14. Deliberately ties the Shuttle design to the hypersonic aircraft named in the title. Each aircraft is profiled in an encyclopedia format.

Becker, John V. "The Development of Winged Reentry Vehicles, 1952-1963." Unpublished study, 57 pp. Copy in NASA History Office Reference Collection. An interesting discussion of the search for a fully reusable spacecraft.

Blaine, J.C.D. *The End of an Era in Space Exploration*. San Diego, CA: American Astronautical Society, 1976. Reviews earlier efforts to place men in space and has some discussion of the Shuttle program.

Bono, Philip. "The Rombus Concept." *Astronautics & Aeronautics*. January 1964, pp. 28-34. One of several articles in this issue devoted to the possibility of a reusable transportation system between the earth and the moon.

Bono, Philip, and Gatland, Kenneth. *Frontiers of Space*. New York: Macmillan Pub. Co., 1969. Profiles efforts to explore space, as well as several proposed methods and objectives for future space activities including a lunar base and a space station. There is a lengthy discussion in the book of the Space Shuttle, as well as its precursors, emphasizing its reusable nature and how it would lower launch costs. There is also a commentary on the 1968 discussion of George Mueller, NASA's Manned Space Flight head, before the British Interplanetary Society about NASA's plans for a Space Shuttle. The commentary is oriented toward the fully-reusable concept, NASA's preference at that time. The narrative is especially interesting for its representation of what was thought about the Shuttle in 1969.

Ehricke, K.A., and D'Vincent, F. "The Nexus Concept." *Astronautics & Aeronautics*. 3 (January 1964): 18-26. One of several articles in this issue devoted to exploring the possibility of a reusable transportation system between the earth and the moon.

Hallion, Richard P., ed. *The Hypersonic Revolution: Eight Case Studies in the History of Hypersonic Technology*. Wright-Patterson Air Force Base, OH: Aeronautical Systems Division Special Staff Office, 1987. 2 Vols. This two volume work contains studies of eight hypersonic R&D programs: the X-15, the X-20A Dyna-Soar, winged reentry vehicles, ASSET, Project PRIME, the Scramjet, lifting bodies, and the Space Shuttle. This study was done to provide information on the evolution of hypersonic technology to program personnel working on the National Aero-space Plane (NASP). It places the Shuttle in the context of what has gone before and suggests that it is just one program in a continuum that led toward the present state of technology, the NASP. The studies in this two-volume set were written over a lengthy period from the 1960s to the present. Some were completed by Air Force historians as internal publications, while others were done in NASA. Hallion pulled them together, and in many cases rewrote sections of them. It is a most detailed, academic, and useful presentation of information.

Hallion, Richard P. "Lifting Bodies: These 'Flying Fish' Were the Forebears of Today's Space Shuttle." *Air & Space*. March-April 1980, pp. 6-7. This short article, written by a leading historian of aeronautical technology, deals with the study, testing and evaluation of lifting bodies during the 1950s and 1960s. With an acknowledgment extended to Eugen Sänger, Hallion briefly describes the development of such programs as Dyna-Soar and other experimental programs. It is oriented toward hardware, and emphasizes the holdings in the National Air and Space Museum of many of these various types of craft.

Hallion, Richard P. *The Path to the Space Shuttle: The Evolution of Lifting Reentry Technology*. Edwards AFB, CA: Air Force Flight Test Center History Office, 1983. An outstanding monograph by one of the leading historians of aviation technology, this study emphasizes the evolution of technology toward the development of a reusable spacecraft. It describes the evolution of the reusable spacecraft concept, emphasizing the work of Eugen Sänger, lifting body studies, and the technological breakthroughs that allowed the Shuttle to be built.

Hallion, Richard P. "The Path to Space Shuttle: The Evolution of Lifting Reentry Technology." *Journal of the British Interplanetary Society*. 30 (December 1983): 523-41. This is a shortened version of Hallion's 1983 monograph by the same title. It describes and shows the evolution of the reusable spacecraft concept, emphasizing the work of Eugen Sänger, the lifting body studies, and the technological breakthroughs that allowed the Shuttle to be built. It is an especially important article because it shows how the technological problems solved in one program were incorporated into the beginnings of the next attempt.

Hallion, Richard P. "The Space Shuttle's Family Tree." *Air & Space*. April-May 1991, pp. 44-46. This short article, taken from Hallion's discussion of the early history of the Shuttle published in *The Hypersonic Revolution*, deals with hundreds of paper studies, experiments, and a handful of aircraft that actually flew and were the antecedents of the Shuttle. It traces the general design of the Shuttle from lifting body technology to the actual configuration that was built and launched in 1981.

Kocivar, Ben. "Flying NASA's X-24B Lifting Body." *Popular Science*. November 1973, pp. 87-91. Short but interesting article on one of the precursors of the Shuttle in the 1960s.

Koelle, H.H., and Rutland, C.H. "Toward a Reusable Earth-Moon Transportation System." *Astronautics & Aeronautics*. 3 (January 1964): 14-17. One of several articles in this issue devoted to exploring the possibility of a reusable transportation system between the earth and the moon. The others in the issue detail different concepts for such an effort.

Lore, Eugene S. "Manned Lifting Entry." *Astronautics & Aeronautics*. May 1966, pp. 54-64. This is a technical article on the potential of spacecraft that could be flown like an airplane. It identified many of the characteristics that were later incorporated into the Shuttle.

Moise, J.C., Henry, C.S., and Swanson, R.S. "The Astroplane Concept." *Astronautics & Aeronautics*. January 1964, pp. 35-40. One of several articles in this issue devoted to exploring the possibility of a reusable transportation system between the earth and the moon.

Office of Manned Space Flight. *NASA's Manned Space Flight Program*. Washington, DC: National Aeronautics and Space Administration, 29 April 1969. Describes the efforts of NASA to place men in orbit and on the moon. It also discusses the next phase of manned flight, the development of a reusable spacecraft for movement of people and supplies to and from orbit.

Peebles, Curtis. "On Wings Into Space." *Spaceflight*. 28 (June 1986): 276-80. A rather general article on the development of the idea and then the technology for a reusable spacecraft. It has a lengthy discussion of 1930s and 1940s research in such aircraft as the X-1. There is also mention of the X-20A Dyna-Soar program, which had many of the same aspirations as the Shuttle program. Finally, there is a discussion of the lifting body studies and the development of the Shuttle concept in the 1960s and 1970s.

Peebles, Curtis. "Project Bomi." *Spaceflight*. 22 (July-August 1980): 270-72. Interesting story of a program by Bell Aircraft to build a reusable spacecraft. This project tried to develop a booster that could lift a Shuttle-type orbiter into space. It was undertaken in the 1950s but abandoned by Bell because it could not undertake an R&D effort of this magnitude without government sponsorship; the company withdrew to concentrate on more immediately commercial prospects.

Peebles, Curtis. "The Origins of the U.S. Space Shuttle-1." *Spaceflight*. 21 (November 1979): 435-42. The first of two articles, this one discusses some of the early scientific work on reusable spacecraft, especially the work of Eugen Sänger. There is also a lengthy description of Project Bomi, the Bell Aircraft Co. program to develop a large delta-winged reusable spacecraft during the 1950s. Peebles also pays close attention to Dyna-Soar, the ill-fated Air Force program to build a reusable vehicle. Finally, there is a discussion of lifting body technology in the 1960s, including the ASSET and M2-F1 programs.

Peebles, Curtis. "The Origins of the U.S. Space Shuttle-2." *Spaceflight*. 21 (December 1979): 487-92. This article follows on from the previous one and extends the discussion of the Shuttle's development from the lifting body studies of the 1960s through the development of the NASA Shuttle. Major components of this effort were the X-23, X-24, and HL-10 research craft.

Phillips, William H. "Flying Qualities from Early Airplanes to the Space Shuttle." *Journal of Guidance, Control, and Dynamics*. 12 (July-August 1989): 449-59. Originally published as an AIAA paper, this is a fine discussion of the development of aircraft technology, firmly concluding that the Shuttle orbiter is a hybrid, the first spacecraft with aerodynamic lift characteristics.

Plattner, C.M. "NASA to Begin Unmanned Tests of New Type of Lifting Shape for Hypersonic Maneuvers." *Aviation Week & Space Technology*. 29 September 1969, pp. 52-58. This news story discusses in detail early plans for a reusable spacecraft with many of the same features as the Shuttle.

Sänger-Bredt, Irene. "The Silver Bird Story." *Spaceflight*. 15 (May 1973): 166-81. This is an article about early, 1930s-1950s, engineering work done by Austrian aerospace designer Eugen Sänger (1905-1961) on a winged means of escaping the atmosphere along the lines of the Space Shuttle. The vehicle was especially to be used as the first stage of booster rockets or to ferry, supply, and furnish rescue equipment for space stations. The basic concepts of the Shuttle, a cross between a powered booster rocket and an aerodynamic glider are presented. The article was written by his former student, co-worker, and wife.

Sänger-Bredt, Irene. "The Silver Bird Story: A Memoir." in Hall, R. Cargill, ed. *Essays on the History of Rocketry and Astronautics: Proceedings of the Third Through the Sixth History Symposia of the International Academy of Astronautics*. vol. 1. NASA Conference Publication 2014, 1977. Examines Eugen Sänger's vision of a "Silver Bird" rocketplane designed in the 1930s to be used as the first stage of booster rockets or to transport rescue equipment for manned space stations. The article continues with a discussion of Sänger's life and research work until his death in 1961. It is written by his collaborator and wife, Irene Sänger-Bredt.

Von Braun, Wernher, and Ryan, C. "Can We Get To Mars?" *Colliers*. 30 April 1954, pp. 22-29. During the Second World War German scientists, including Wernher von Braun, began testing spacecraft models based on Sänger's concepts as well as theories of their own. This article popularized the idea of a reusable earth-to-orbit space transportation system.

Wilkinson, Stephan. "The Legacy of the Lifting Body." *Air & Space*. April/May 1991, pp. 51-62. Solid, popularly written article which deals with a precursor of the Space Shuttle. Lifting bodies were first tested at Dryden Flight Test Facility in 1963 and served to provide data for the design of the Shuttle. This article details the interesting story of how they were developed on a shoestring and without NASA headquarters approval.

Winter, Frank. "1928-1929 Forerunners of the Shuttle: The 'Von Opel Flights'." *Spaceflight*. 21 (February 1979): 75-83, 92. Well-done story of the pioneering efforts to design and build reusable spacecraft.

Link to Part 2 (1992–2011), Chapter 2—Precursors

CHAPTER 3
THE SHUTTLE DECISION

America's Next Decades in Space: A Report of the Space Task Group. Washington, DC: National Aeronautics and Space Administration, September 1969. This seminal report published just months after the first moon landing describes NASA's plans for the future. It offers several important recommendations relative to the development of the Shuttle. It emphasizes the need for continued exploration of space and the requirement for economy and reusability of spacecraft. This was couched in terms of supporting a space station for planetary exploration. All of these activities would support practical as well as scientific programs. The plan emphasizes the establishment of a space station by 1976 that would be supported by a Shuttle.

Aspin, Les. "The Space Shuttle: Who Needs It?" *The Washington Monthly*. September 1972, pp. 18-22. The author, a Democratic Congressman from Wisconsin, suggests that the Shuttle was the result of NASA's desire to continue as a separate entity. He notes that while the DOD and HUD are critical components of government, the same is not the case with NASA and space exploration. It is a luxury that can be expended when economic pressures require it. He argues that the agency has lived on public relations, and that Congress has enjoyed this glitter as well. He is skeptical of the necessity of the Shuttle and chalks its support up not to legitimate requirements but to NASA "puffery."

Barfield, Claude. "Space Report/NASA Gambles its Funds, Future on Reusable Space Shuttle Program." *National Journal*. 3 (13 March 1971): 539-51. Discusses efforts by NASA to obtain approval of the Space Transportation System. It describes the conceptualization of the program and its emphasis on practical benefits over national prestige. This was a significant and necessary alteration because of the changes in the national economy and international relations. The author concludes that NASA had no option but to develop the Shuttle if it were to remain a well-funded agency. Barfield asserts that the ten months after this article appeared would be critical to the Shuttle program, as forces lined up on both sides of the issue.

Barfield, Claude. "Technology Report/Intense Debate, Cost Cutting Preceded White House Decision to Back Shuttle." *National Journal*. 4 (12 August 1972): 1289-99. An excellent account of the investigations and controversy preceding the Nixon Administration's endorsement of a new Space Transportation System.

Barfield, Claude. "Technology Report/NASA Broaden's Defense of Space Shuttle to Counter Critic's Attacks." *National Journal*. 4 (19 August 1972): 1323-32. Another well-researched and well-written article on the Shuttle debate in the early 1970s.

"Correcting the Mistakes of the Past: A Conversation with John Logsdon." *Space World* (August 1986): 12-18. In this interview, Logsdon argues that the Shuttle neither guarantees routine access to space, nor is inexpensive. This, says Logsdon, was due to unrealistic expectations about the Shuttle's capabilities. Performance would have been better had the designers concentrated on transporting Shuttle-unique payloads. Logsdon claims that basically the same thing is happening to the Space Station. Other topics covered are the role of the President, the Office of Management and Budget, and Congress in the U.S. space program, and the Soviet space program.

Dooling, Dave. "Space Shuttle: Crisis and Decision." *Spaceflight*. 14 (July 1972): 242-45. This is an interesting article, written very early in the Shuttle program, about the decision to build the spacecraft. It describes some, but not all, of the political machinations involved in the decision.

Draper, Alfred C.; Buck, Melvin L.; and Goesch, William H. "A Delta Shuttle Orbiter." *Astronautics & Aeronautics*. 9 (January 1971): 26-35. This is an excellent technical review of the reasons for developing a delta-wing versus a straight-wing or lifting body orbiter. The authors were engineers for the Air Force Flight Dynamics Laboratory, and their arguments contributed to the decision to change to a delta configuration, giving the military the 2000 mile crossrange capability it needed for military missions.

Economic Analysis of New Space Transportation Systems: Executive Summary. Princeton, NJ: Mathematica, Inc., 1971. This s tudy p resents an eco nomic an alysis o f al ternative s pace t ransportation systems. It indicates that the expendable systems represent modest investments, but the recurring costs of operation remain high. The Space Shuttle and tug system requires a substantial investment but would substantially reduce the recurring costs of operation. Economic benefits and costs of the different systems are also analyzed.

Farrar, D.J. "Space Shuttle and Post Apollo." *Aeronautical Journal*. March 1973. pp. 157-62. This article, written by the Coordinating D irector of P ost-Apollo S tudies f or th e B ritish A ircraft C orporation, is c oncerned w ith t he relationship of the Apollo program to the Shuttle effort and the role of international cooperation concerning the development of the new spacecraft. The author sees all manner of opportunity for British use of the Shuttle and urges close cooperation in the program's execution.

Fletcher, James C. "Are SKYLAB and the Space Shuttle Worth the Investment?" *Government Executive*. January 1974. pp. 38 -40, 4 2. T his is a lo gical d efense o f b oth th e S huttle a nd S kylab p rograms w ritten b y th e NASA Administrator. He justifies th e Shuttle on th e basis of c ost s avings ($1 billion p er year for o perations) and versatility, and its ability to serve as an excellent platform for scientific research, to mate humans and machines in a r eliable a nd m eaningful w ay, to m ix s cientific an d p ractical ap plications, t o p rovide a s pace r escue capability, to be used for D epartment o f D efense p rojects, t o f oster aer ospace t echnology, t o f acilitate international cooperation, and to provide the key to U.S. supremacy in space.

GAO R eport on A nalysis of C ost of Spac e Shut tle P rogram. W ashington, D C: U .S. H ouse o f Representatives Committee on Science and Astronautics, 1973. This is a set of hearings held by the House subcommittee on Manned Space Flight on 26 June 1973 concerning the GAO report on the feasibility of the Shuttle versus two alternative methods of launch operation.

General Accounting Office. *Analysis of Cost Estimates for the Space Shuttle and Two Alternate Programs*. Washington, DC: General Accounting Office, 1973. This report reviews the costs associated with the use of a fully reusable Shuttle, a partially reusable Shuttle, and a fully expendable launch system. It finds that the partially reusable Shuttle is cost effective with a rigorous flight schedule over several years.

General Accounting Office. *Cost Benefit Analysis Used in Support of the Space Shuttle Program*. Washington, DC: General Accounting Office, 1 972. T his 5 3-page report describes the process by which N ASA developed its Shuttle c ost-effectiveness ar gument. This analysis w as s een as a factor in the presidential decision to p ress ahead with development of the spacecraft.

Gibson, T.A. and Merz, C.M. *Impact of the Space Shuttle Program on the Economy of Southern California*. Space Division, North American Rockwell, SD 71-7662, September 1971. Discusses the probable impact on regional employment a nd p roduction i n S outhern C alifornia t hrough t he a warding o f p rime c ontracts for the Space Shuttle p rogram. T he p aper concludes that t his r egion, w ith a h eavy co ncentration o f aerospace i ndustries, would find a "highly favorable and widely diffused" economic impact from such contracts.

Gillette, Robert. "Space Shuttle: Compromise Version Still Faces Opposition." *Science*. 175 (28 January 1972): 392-96. Reviews the controversy surrounding the decision to build the Shuttle, especially the configuration debate that took place within the government before the presidential announcement of 5 January 1972. Gillette also notes that in spite of presidential support, the Shuttle had its share of critics in Congress and that it could be tabled by the l egislative b ranch. F or o ne, C ongress q uestioned the argument for the cost-effectiveness of t he Shuttle. Gillette is skeptical of the Shuttle and calls the program "NASA's ferryboat to the future."

Guilmartin, John F., Jr., and Mauer, John Walker. *A Shuttle Chronology, 1964-1973*. Houston, TX: Lyndon B. Johnson

Space Center, 1988. 5 Volumes. This is a comprehensive and detailed chronology of the development of the Shuttle, d ivided in to separate sections co ncerning v arious as pects o f t he p rogram an d o rganized chronologically within them. It suffers from some repetition, but still has much valuable information and many reference notes.

Hechler, K en. *Toward t he E ndless F rontier: H istory of t he C ommittee on Sc ience and T echnology, 1959-1979* (Washington, DC: U.S. House of Representatives, 1980). Contains the best account to date of Congressional wrangling over the Shuttle, and demonstrates the bipartisan nature of both Shuttle support and opposition.

Heiss, K laus P., and Morgenstern, Oskar. *Mathematica Economic Analysis of the Space Shuttle System*. Princeton, NJ: Mathematica, Inc., 1972. This is a t hree volume study of the economic value of the Shuttle. It found that the major economic potential for the Shuttle in the 1980s would be the lowering of space program costs due to the reuse, refurbishment, and updating of satellite payloads. This is based on a partially reusable, stage-and-a-half Shuttle. It uses sophisticated statistical models to show the measure of economic viability of the system.

Heiss, K.P. "Our R and D Economics and the Space Shuttle." *Astronautics & Aeronautics*. 9 (October 1972): 50-62. This lengthy article is an economic analysis of the effect of R&D projects on the Shuttle.

Holden, Constance. "Space Shuttle: Despite Doubters, Project Will Probably Fly." *Science*. 180 (27 April 1973): 395, 397. This short article deals with the debate in Congress over the NASA budget for FY 1974, especially as it relates t o t he f unding t o b e e xpended on the Shuttle. It d escribes the efforts of cr itics, am ong t hem space scientists who saw the Shuttle as eating into their programs, to kill or at least delay the program. There is an explicit tie between the Shuttle debate and that surrounding the recently cancelled Supersonic Transport, with the author asserting that the two programs are comparable.

Holder, William G., and Siuru, William D., Jr. "Some Thoughts on Reusable Launch Vehicles." *Air University Review*. 22 (November-December 1970): 51-58. Discusses one of the central problems that led to the development of the Shuttle, the quest for cost-efficient launch vehicles through the development of reusable systems. It sets the stage for the Shuttle debate and decision.

Hotz, Robert. "The Shuttle Decision." *Aviation Week & Space Technology*. 31 July 1972. p. 7. This editorial applauded the 5 January 1972 decision of President Richard Nixon to proceed with the development of the Space Shuttle. It does not analyze how the decision was made so much as cheer the nation's commitment to leadership in the space age.

Howell, Craig. "The S huttle W alks a T ightrope." *New S cientist*. 59 (9 A ugust 1973): 321-23. R eviews th e h istory of budget d ifficulties f or th e d evelopment o f th e S pace S huttle. T he a uthor b elieves that if the B ureau of the Budget does not supply NASA with the full funding needed to maintain the space program at its present level, NASA will sacrifice whatever is necessary to k eep th e Shuttle going. Suggests that it is important that th e Shuttle p rogram be funded at close to the optimum rate because speeding up or slowing down from that rate will increase costs. The consequences of a significant cost overrun or a suspension of the Shuttle are examined, with the author predicting serious repercussions.

Hunter, Max well W ., I I, Mi ller, W ayne F ., an d G ray, R obert M . " The S pace S huttle W ill Cut Payload Costs." *Astronautics & Aeronautics*. 10 (June 1972): 50-58. This article, written by three engineers from the Lockheed Missiles and Space Co., argues that while the Shuttle is not being designed solely for the purpose of reducing operating costs, it does take advantage of technological developments during the first decade of spaceflight to achieve a quantum leap in the capabilities of the spacecraft. The article discusses at length the parameters of the effort to d evelop the S huttle, analyzing c osts and e stimating p ossible returns. The result, the authors believe, would be a great improvement over earlier space operations as well as a reduction in operating costs.

Kyger, Timothy B. "Lunar Eclipse: President Nixon, Public Opinion, and Post-Apollo Planning." Bachelors of Public Administration Thesis, University of San Francisco, 1985. This thesis attempts to determine to what extent public opinion influenced Nixon's decision to severely curtail NASA's space program. It concludes that adverse public opinion in 1969 about the future of the space program weakened congressional support for the recommendations of the Space Task Group, which in turn lessened Nixon's own endorsement of these recommendations.

Layton, J. Preston. "Our Next Steps in Space: A Status Report on New Space Transportation Systems." *Astronautics & Aeronautics*. 10 (May 1972): 56-65. Because of the efforts of NASA to develop the Space Shuttle the AIAA convened an ad hoc panel to assess the new space transportation system. This article reviews the efforts of the nation to build the Shuttle up to this early date, describing some of the various concepts and tracing the chronology of the Shuttle decision.

Levine, Arthur L. *The Future of the U.S. Space Program*. New York: Praeger Publishers, 1975. Chapters six and seven of this book deal almost exclusively with the post-Apollo space policy struggle and contain good descriptions and a useful early analysis of the emergence of the Shuttle as *the* NASA program of the 1970s.

Logsdon, John M. "The Decision to Develop the Space Shuttle." *Space Policy*. 2 (May 1986): 103-19. Surveys the policymaking process within government on the Shuttle program. In response to the *Challenger* accident Logsdon asserted that the Space Shuttle decision essentially set up the program for a disaster. He reviews the decision, announced on 5 January 1972, to develop a specific Shuttle design. Logsdon believes it was a bare-bones funding strategy for the program and chides the bureaucracy for politicizing the process. This decision was influential in NASA's inability to deliver routine and inexpensive space transportation. According to NASA plans in 1969, the Shuttle was to consist of two reusable components. After launch, the booster stage would be flown by its crew to a landing near its launch site, while the orbiter would continue on into space. As a result of budgetary restrictions, Logsdon asserts, these plans had to be abandoned. The result, after very extended evaluations and negotiations, was the Shuttle design in its current form, which was characterized by smaller development costs but substantially larger operating costs. This article is very similar to Logsdon's other studies on the subject and reflects on the *Challenger* disaster in relation to the policy decisions over the life of the Shuttle program.

Logsdon, John M. "From Apollo to Shuttle: Policy Making in the Post Apollo Era." Unpublished partial manuscript, Spring 1983, copy in NASA History Office Reference Collection. This is a detailed and insightful study of the political process involved in the decision to build the Space Shuttle in the late 1960s and early 1970s. It represents a more detailed discussion of the same subject, with essentially the same conclusions, that Logsdon presented in his articles on the Shuttle.

Logsdon, John M. "Shall We Build the Space Shuttle?" *Technology Review*. October-November 1971. The author, one of the leading analysts of space policy, prepared this article at the same time that NASA was trying to win approval of the Shuttle program from the Nixon administration. Logsdon reviews the issues at play in Washington in 1970-1971 and how they affected the funding question. He finds that the only comparably-sized space program, Apollo, operated in an environment in which political and economic decisions were strikingly different from those affecting the Shuttle. A key point was that presidential support for Apollo was omnipresent and cast an overarching shadow on all policy issues. Such was not the case for the Shuttle; support for it was at best ambivalent and at some extremes perhaps contentious. Logsdon also contends that the political process, with officeholders constantly seeking popular support and reelection every 2, 4, or 6 years, means that they want payoffs in their programs within those time constraints. The process is ill-suited to fostering long-term technological programs with results only coming in future decades.

Logsdon, John M. "The Space Shuttle Decision: Technology and Political Choice." *Journal of Contemporary Business*. 7 (1978): 13-30. In a detailed, scholarly, and convincing article, Logsdon reviews the policymaking process for the Shuttle. His principle conclusion is that "the Shuttle was approved, as a means of operating in space, *without any extensive debate over what the goals of space operations in the 1980s might be*" (p. 27). While much debate over technological designs and Shuttle configurations resulted from the process, the requirement for its operation was not firmly established. The author suggests that one of the strengths of the American system is that there is give and take over issues and pragmatic compromise to achieve results that are acceptable to the widest range of viewpoints, but that in the heavily technological arena it is of questionable virtue. The result was the development of a Shuttle that might not meet the needs of the nation.

Logsdon, John M. "The Space Shuttle Program: A Policy Failure." *Science.* 232 (30 May 1986): 1099-1105. In a thoughtful article, Logsdon contends that the decision to build the Shuttle emerged from a murky policymaking process that did not properly analyze the approach or gauge the operational capability and, more importantly, compromised the funding levels so badly that serious technological compromises resulted as well. He notes that NASA allowed its Shuttle hopes to be held hostage by political and economic forces. The program gained its support on a cost-effective basis, rather than on scientific, technological, or other grounds. This ensured that the budget-cutters would hack away at the program every year. It also suffered from a lack of strong support from key political figures. There were no Kennedys or Johnsons to champion the Shuttle and the result was a politicization of the process and what Logsdon calls a "policy failure."

Mathews, Charles W. "The Space Shuttle and its Uses." *Aeronautical Journal.* 76 (January 1972): 19-25. This article assesses the development of a reusable Shuttle system, noting that it was made practicable by the availability of improved, staged combustion engines and durable, thermal protection systems. The two stage launch configuration with fully-reusable boosters and orbiter elements is considered to be the best design solution, and size specifications for such a vehicle are examined as a function of launch costs. Significant vehicle characteristics are explained in terms of cargo bay dimensions, cross-range maneuvering capability, mission duration requirements, engine characteristics, and acceleration constraints. The Shuttle flight activities that Mathews foresees include satellite deployment and recovery, research, and space station support operations. Phases of the development program are also outlined, and structural details of several candidate Space Shuttle concepts are illustrated.

Merz, C.M., Gibson, T.A., and Seitz, C. Ward. *Impact of the Space Shuttle Program on the National Economy.* Space Division, North American Rockwell, SD 71-478, March 1971. This paper reports on the results of a study to determine the overall impact of an $8.6 billion Space Shuttle program on the national economy. Factors examined were the value of production by industry, the amount of employment by industry, and the effect of foreign trade. The estimated impact on these areas was then compared with the probable impact caused by similar expenditures in residential construction and consumer spending. The study concludes that the economic impact of the Space Shuttle program compares very favorably with the benefits stemming from similar investments in residential construction or an increase in consumer spending.

Mueller, George E. "The New Future For Manned Spacecraft Developments." *Astronautics and Aeronautics.* 7 (March 1969): 24-32. Argues for the advantages to be gained by placing a permanent space station in orbit, and the necessity of building a Shuttle to transport materials to the Space Station at low cost, by the NASA head of spaceflight. The Shuttle envisioned by Mueller has many of the advantages of a commercial airplane and would be able to carry payloads into orbit at the cost of about $5 per pound.

Myers, Dale D. "The Shuttle: A Balancing of Design and Politics." *Issues in Program Management*, Summer 1992, pp. 42-45. Analyzes the various cost considerations that influenced the decision to build the Shuttle. Lack of adequate operational models and overly optimistic cost-effectiveness estimates characterized the early planning stages. This and the emphasis of development over operation were major causes of the Shuttle's later problems, the article asserts.

"NASA in Trouble with Congress, Executive, Scientists." *Nature.* 231 (11 June 1971): 346-48. Describes the difficulties between NASA and the other branches of government over the funding of the Shuttle, delineating very well the differences between the Apollo program, which had a Presidential mandate, and the Shuttle, which had reluctant support at best. Concludes that only because of DOD involvement did the Shuttle gain sufficient support to go forward. It ends: "In staking its future to the Shuttle, perhaps a necessary move, NASA had made a devil's pact with the military, ignored the advice of the scientific community and risked antagonizing its supporters in Congress by sacrificing peripheral projects. The gamble is a dangerous one, but at least if it fails NASA will end with a bang, not a whimper" (p. 348).

NASA Space Shuttle Summary Report. Washington, DC: National Aeronautics and Space Administration, rev. ed., 31 July 1969. Summary report of efforts to define the proposed Space Shuttle. It should be used in conjunction with the longer 4-volume report described immediately below.

NASA Space Shuttle Task Group Report. Washington, DC: National Aeronautics and Space Administration, 1969. 4 Volumes. This multi-volume report, written by a joint NASA/DOD study group, was the foundation of NASA's early efforts to define the Space Shuttle program. Volume one contains the summary and makes a strong case for the development of "A versatile Space Shuttle system that can transport effectively, a varying mix of personnel and cargo to low earth orbits and return, could be the keystone to the success and growth of future space flight developments for the exploration and beneficial uses of near and far space." The second volume deals with "Desired System Characteristics," the third with "Vehicle Configurations," and the fourth with "Program Plans." Originally issued on 19 May 1969, it was revised and reissued on 12 June 1969. It projected the first operational flight of a Shuttle by 1980, and proposed three options for a Shuttle: fully-reusable, one-and-one-half stage or drop tank concepts, and expendable boosters plus reusable orbiter.

NASA Space Task Group. *Technology Program Plan.* Washington, DC: National Aeronautics and Space Administration, 1969. This report, issued on 26 June 1969, is an outgrowth of Space Task Group studies and emphasizes five areas: aerodynamics/configuration selection, integrated electronics system, expendable tank construction, propulsion, and thermal protection. This plan contains specifics of development for these five areas since they are critical for the timely development of the system. Subcommittees in each of the five areas develop action plans for technology harnessing that are laid out in this work.

NASA Space Task Group. *The Post Apollo Space Program: Directions for the Future.* Washington, DC: Government Printing Office, September 1969. A seminal document in the development of the Shuttle, this work analyzes of the possibilities for the development of a reusable spacecraft at the time that NASA is seeking a follow-on program for the lunar expeditions. It was generated by the presidentially appointed group considering the best direction for the U.S. space program after the Apollo program. It recommends a goal of a balanced human space flight and science with five policy goals: (1) expand the space applications program to realize potential benefits, (2) enhance the defense posture of the U.S. through the exploitation of space techniques for military missions, (3) increase knowledge of the universe through a strong program of lunar and planetary exploration, astronomy, physics, and earth and life sciences, (4) develop new systems and technology for space with emphasis on reusability, commonality, and economy, and (5) promote a sense of world community through a program providing opportunity for broad international participation.

The Next Decade in Space: A Report of the Space Science and Technology Panel of the President's Science Advisory Committee. Washington, DC: President's Science Advisory Committee, March 1970. This important report reviews the development of the space program in the United States through the moon landing and projects some future objectives for the President. There is some discussion of the space transportation system and the report concludes: "A Space Shuttle will allow large payloads to be assembled in orbit, with consequent advantages for manned flight. This may provide the line of evolution towards systems for long-duration flights or to a greater variety of manned activity should this be desirable" (p. 37). It also notes that the development of reusable systems is critical to lowering costs for orbit. It recommends that NASA continue efforts to develop the Shuttle and aim for a decision on it by fiscal year 1972. It does not support NASA's two other post-Apollo goals: a human mission to Mars and a space station.

O'Leary, Brian. "The Space Shuttle: NASA's White Elephant in the Sky." *Bulletin of Atomic Scientists*. 29 (February 1973): 36-43. This essay is highly critical essay of the Space Shuttle. The author notes that the principal problems with the Shuttle include its questionable role in competing national priorities; the lack of a clear definition of NASA's goals for the Shuttle; the uncertainties of the recurring costs; the question of payload subsystem refurbishments; the secondary social and economic costs of the Shuttle; and the probability that the Department of Defense will become the primary user of the Shuttle and therefore drive the configuration and costs. O'Leary based his arguments on an oral presentation he made before the Senate Committee on Aeronautical and Space Sciences when it was considering the 1973 budget. He summarized the skepticism of many within the scientific community. The article is an interesting critique but contains no scholarly references.

Pace, Scott. "Engineering Design and Political Choice: The Space Shuttle, 1969-1972." M.S. Thesis, MIT, May 1982. Detailed academic study of the interplay of engineering design and political factors during the early stages of Shuttle development. During this period the concept went from a rocketplane take-off and landing proposal to the take-off-like-a-rocket and land-like-a-glider system that was developed. Although interesting, it is nonetheless a thesis and not the product of a mature historian.

Redford, Emmette, and White, Orion F. *What Manned Space Program After Reaching the Moon? Government Attempts to Decide, 1962-1968*. Syracuse, NY: The Inter-University Case Program, January 1971. Limited edition study of the efforts of NASA and other government agencies to determine what policies and programs it should pursue for the future space program. It is especially helpful as a statement of where leaders thought the U.S. should be going at the very time the debate over the development of the Shuttle was taking place.

"Reusable Space Shuttle Effort Gains Momentum." *Aviation Week & Space Technology*. 27 October 1969, pp. 22-24. Useful article describing the efforts up to that point to "sell" the idea of the Shuttle based on its cost efficiencies.

Selection of Papers Presented at the Space Shuttle Symposium, Smithsonian Museum of Natural History, Washington, DC, October 16-19, 1969. (Washington, DC: National Aeronautics and Space Administration, 1969). An exceptionally important collection of papers that discusses the origins, development, and promise of a reusable Shuttle for entrance into and recovery from earth orbit. Individual essays deal with a wide variety of subjects, everything from the development of launch vehicles to the potential for a space station.

Shaver, R.D., Dreyfuss, D.J., Gosch, W.D., and Levenson, G.S. *The Space Shuttle as an Element in the National Space Program*. Santa Monica, CA: Rand Corporation, October 1970. This report for the United States Air Force assesses the role of the proposed Space Shuttle in policy and technological issues. It especially examines the economic justification and potential funding problems of the Shuttle as advanced by NASA to the President in September 1969. It suggests that the concept of a two-stage, fully-reusable launch vehicle that can place a 40,000- to 50,000-pound payload in polar orbit would show a net savings of $2.8 billion by 1990. To achieve this, however, the government would have to fund NASA at the peak of $7 billion in 1975, about double NASA's 1970 budget. The authors conclude that viewed over the long term, the Shuttle had definite merit, but its immediate economic justification depended on the pace that was finally adopted for the national space program.

"Space Shuttle: NASA Versus Domestic Priorities." *Congressional Quarterly*. 26 February 1972, pp. 435-39. This short article discusses the issues relative to the political football of the Shuttle. It quotes at length the pros and cons of the system from key Congressional and Executive Branch personnel as they understood the issue at the time of Nixon's approval of the Shuttle.

Truax, Robert C. "Shuttles—What Price Elegance?" *Astronautics & Aeronautics*. 8 (June 1970): 22-23. This is an important "minority report" on the Shuttle's modus operandi as it was being designed and before it was approved as a program by President Nixon. It argues that the necessity of a fully-reusable Shuttle is a chimera. Traux argues for an expendable or partially reusable lower stage and a reusable orbiter, but he contends that there was no necessity of making it a winged or lifting body vehicle. Instead, a ballistic craft would do just as well and be recoverable in the ocean and reusable. That would cut down development costs drastically, but since splash-downs were "inelegant" NASA was committed to a winged spacecraft that "could be an unparalleled money sponge."

Link to Part 2 (1992–2011), Chapter 3—The Decision to Build the Space Shuttle

CHAPTER 4
SHUTTLE DESIGN AND DEVELOPMENT

Bailey, R.A., and Kelley, D.L. " Potential of Recoverable Booster Systems for Orbital Logistics." *Astronautics & Aeronautics.* 2 (January 1964): 35-40. One of several articles in this issue devoted to exploring the possibility of a reusable transportation system between the earth and the moon.

Baker, David. "Evolution of the Space Shuttle, Part 1." *Spaceflight.* 15 (June 1973): 202-10. First in a series of articles dealing with the development of the Shuttle. It focuses on the technological development of the system through the proposed baseline for Phase B studies.

Baker, David. "Evolution of the Space Shuttle, North American Rockwell, Part 2." *Spaceflight.* 15 (July 1973): 264-68. Second in series of articles, this one reviews the studies in the late 1960s on the possibility of a fully-reusable Shuttle, emphasizing technological development.

Baker, David. "Evolution of the Space Shuttle, North American Rockwell, Part 3." *Spaceflight.* 15 (September 1973): 344-52. The third Shuttle article in this series deals with the technological development of the external fuel tanks for the orbiter and the replacement of the piloted flyback booster with a ballistic booster arrangement.

Baker, David. " Evolution of the Space Shuttle, Part 4 ." *Spaceflight.* 18 (September 1976) : 304 -38. Written to commemorate the roll-out of the Orbiter 101 prototype, this article reviews the major design changes and presents an evolutionary update of the Shuttle's technological development.

Baker, David. "Evolution of the Space Shuttle: 7. External Tank Design-1." *Spaceflight.* 20 (February 1978): 60-66. An article relating the basic development of the technology going into the external tank design.

Baker, David. "Evolution of the Space Shuttle: 8. External Tank Design-2." *Spaceflight.* 20 (March 1978): 110-15. A continuation of the above article.

Beichel, Rudi. " Nozzle Concepts for Single-Stage Shuttles." *Astronautics & Aeronautics.* 13 (June 1975) : 16 -27. Technical article on breakthroughs associated with the propulsion systems being developed to launch the Shuttle orbiter into space.

Bekey, I., and Mayer, H. "1980-2000: Raising Our Sights for Advanced Space Systems." *Astronautics and Aeronautics.* 14 (July-August 1976): 34-64. This special section on the possible future of space travel has much to say about space stations and flights to Mars and beyond, but also emphasizes the development of the Shuttle as the necessary first step in any ready access to space.

Bell, M.W. Jack. "Advanced Launch Vehicle Systems and Technology." *Spaceflight.* 20 (April 1978): 135-43. This article is a good report on the development of the launch vehicle that would be used to send the Shuttle orbiter into space.

Bourland, C.T., Rapp, R.M., and Smith, M.C., Jr. "Space Shuttle Food System." *Food Technology.* 31 (September 1977): 40-41, 44-45. This is a general article on the food system being developed for use in orbit.

Brown, Nelson E. " Safe Shuttle." *Technology Review.* 79 (March/April 1977) : 17 -25. This is an early study of the redundant systems and other safety features being incorporated into the Shuttle design by a leading investigator of safety programs.

Brown, Nelson E. "Space Shuttle's Safety and Rescue: An Enormous Jump In Man's Ability to Work in Outer Space."

Space World. 10 (December 1977): 16-25. This is a complex article that analyzes the ability of the Shuttle to ensure that no individual is stranded in space and to teach us how best to live in a space environment.

Bursnall, W.J.; Morgenthaler, G.W.; and Simonson, G.E., eds. *Space Shuttle Missions of the 80's*. San Diego, CA: Univelt Inc., 1977. A multi-author review of the possibilities offered for missions by the Shuttle. It contains the proceedings of the 21st annual meeting of the American Astronautical Society in Denver, CO, 26-28 August 1976.

Caveny, Leonard H. "Thrust and Ignition Transients of the Space Shuttle Solid Rocket Motor." *Journal of Spacecraft and Rockets*. 17 (November/December 1980): 489-94. This is a technical article on the solid rocket propulsion system developed by Morton Thiokol for the Shuttle.

Chaffee, Norman, comp. *Space Shuttle Technical Conference Papers*. Washington, DC: National Aeronautics and Space Administration, 1985. This two volume work contains papers on many aspects of the design and performance of the Shuttle. It was the product of a conference held at the Johnson Space Center on 28-30 June 1983.

Collingridge, David. "Technology Organizations and Incrementalism: the Space Shuttle." *Technology Analysis and Strategic Management*. vol. 2, no. 2, 1990: 181-200. This article argues that the Shuttle's performance has been poor because it was built using inflexible technology. This was a result of technology development through a centralized process dominated by a few very similar organizations with little debate or compromise, and with risks being taken at the expense of the tax payer. The authors conclude that better technological performance would result from incremental development and decentralized decision making.

Cooper, A.E., and Chow, W.T. "Development of On-Board Space Computer Systems." *IBM Journal of Research and Development*. 20 (January 1976): 5-19. This is a scholarly article discussing the design and construction of a completely new on-board computer system for the Shuttle. This effort spawned much of the technology now present in modern microcomputers, especially the micro-chip.

Cooper, Paul, and Holloway, Paul F. "The Shuttle Tile Story." *Astronautics & Aeronautics*. 19 (January 1981), pp. 24-34, 36. This is a good discussion of the development of the special tiles used to protect the Shuttle during reentry. While it contains some technical information about the tiles, it is oriented toward a general audience.

Dankoff, Walter; Herr, Paul; and McIlwain, Melvin C. "Space Shuttle Main Engine (SSME)—The `Maturing' Process." *Astronautics & Aeronautics*. 21 (January 1983): 26-32, 49. This technical article describes the origins and evolution of the Shuttle's main engine. It seeks to show that the main engine, which had received some negative press, was a well-designed and efficient component of the Shuttle. It suggests that the SSME had proven itself "in hundreds of ground tests and five flights of *Columbia*." Even so, it had been repeatedly redesigned and improved with every orbiter built for the program.

"Designing for Zero-G: The Space Shuttle Galley." *Design News*. 22 October 1979, pp. 48-50. Although not a particularly scintillating topic, this article treats something not usually discussed in relation to the Shuttle, the galley of the spacecraft and the difficulties of preparing foods in microgravity. The emphasis of this piece is toward the specially-designed and built equipment of the galley.

Donlan, Charles J. "Space Shuttle Systems Definition Evolution." *Issues in Program Management*, Summer 1992, pp. 46-48. This article reviews some of the Shuttle configurations considered in the early 1970s, stressing their appeal from a cost-effectiveness standpoint. The discussion concentrates on the development of the booster rockets.

Dooling, Dave. *Shuttle to the Next Space Age*. Huntsville, AL: Alabama Space and Rocket Center, 1979. A collection of presentations from the Alabama Section of the American Institute of Aeronautics and Astronautics, this book contains 22 articles on a variety of subjects related to the Shuttle. Organized in several sections—National Space Line, Space Applications, Space Science, and Other Space Activities—These papers are largely technical and designed for an academic audience. There are no Shuttle program overviews or historically oriented articles in this publication.

Draper, Alfred C.; Buck, Melvin L.; and Goesch, William H. "A Delta Shuttle Orbiter." *Astronautics & Aeronautics*. 9 (January 1971): 26-35. This is an excellent technical review of the reasons for developing a delta-wing versus a straight-wing or lifting body orbiter. The authors were engineers for the Air Force Flight Dynamics Laboratory, and their arguments contributed to the decision to change to a delta configuration, giving the military the 2000 mile crossrange capability it needed for military missions.

Elson, Benjamin M. "Shuttle Booster Motor Tests Planned." *Aviation Week & Space Technology*. 20 February 1978, pp. 54-59. This is a lengthy article on the development and testing of the Shuttle's booster engines.

Faget, Maxime A. "Space Shuttle: A New Configuration." *Astronautics & Aeronautics*. 8 (January 1970): 52-61. This is an exceptionally important article written by one of NASA's foremost engineers that looks at the plans for the development of the Shuttle and offers a configuration for a fully-reusable, straight-wing, two-stage system. It contains considerable technical detail of the Shuttle. Faget concludes that his configuration offers complete reusability, economical cost per flight, and a tremendous advantage to the United States' efforts to make space more accessible.

Farrar, D.J. " The Space Shuttle: Concept and Implications." *Spaceflight*. 14 (March 1972) : 104 -108. This paper describes the Shuttle, tug, and orbital station transportation system envisioned by NASA and assesses their costs and benefits. It notes as especially significant the European economic implications.

Fitzgerald, Paul E., Jr. and Gabris, Edward A. "The Space Shuttle Focused-Technology Program: Lessons Learned." *Astronautics & Aeronautics*. 21 (February 1983): 60-67, 72. This article reviews the technological program for the Shuttle, emphasizing the structure and membership of the steering committee, how it functioned, and how NASA "put wheels under" the technology development program for Shuttle in several arenas. These included: propulsion, electronics, aerothermodynamics, aeroelasticity, materials, and biotechnology. The conclusions aimed toward the use of focused-technology development for cost-avoidance and efficient methodology. NASA had a clear picture of what it wanted and organized a research group from several sites and disciplines to work on pieces of it.

Gatland, Kenneth. "Designing the Space Shuttle." *Spaceflight*. 15 (January 1973): 11-14. Presents information, as it was known at that time, about the configuration of the Shuttle. It describes the orbiter, the reusable boosters, and the expendable liquid fuel tank. It also discusses the Shuttle's avionics systems and the thermal protective effort.

Gatland, Kenneth. " The Space Shuttle." *Spaceflight*. 13 (May 1971) : 158 -63. This article describes the Shuttle as conceived in 1970-1971. It emphasizes the two-stage, fully reusable system with crews in each component of the Shuttle, the booster and the orbiter. There is also considerable technical detail in the article about how the reusable system would operate.

Geddes, J. Philip. " Space Shuttle Basics." *Interavia*. 27 (December 1972): 1331 -34. This article, prepared by a staff writer at *Interavia*, is a good explanation of the Shuttle's mission as understood in 1972, its technological elements, and its challenges for development. Most important, it is one of the earliest full explanations of the Shuttle configuration as eventually built.

General Accounting Office. *Space Shuttle: Changes to the Solid Rocket Motor Contract TLSP: Report to Congressional Requestors*. Washington, DC: General Accounting Office, 1988. This report, done after the *Challenger* accident, describes the changes to the Space Shuttle solid rocket motor contract, and assesses the redesign of the m otors f ollowing t he a ccident, d escribing t he c hanges i n t he m otor j oints a nd other design changes to enhance the motor's safety and reliability. These changes were incorporated into 1 3 sets of boosters for the Shuttle. It also comments on the method used to assess the costs of these changes, noting that the fees paid were changed from specific cost and performance incentives to more subjective valuations by NASA.

General Accounting Office. *Status and Issues Relating to the Space Transportation System*. Washington, DC: General Accounting Office, 21 April 1976. This study assesses NASA's Shuttle development plan and concludes that it could result in increased costs, schedule d elays, an d p erformance d egradation t hat w ere n ot o riginally envisioned. The d evelopment p lan, r evised a s t he p rogram f ell b ehind s chedule a nd t ook f unding c uts, embodied such factors as reduced testing, compressed schedules, and concurrent development and production. The s tudy a lso a sks, b ut d oes not truly answer, w hether the Shuttle s ystem f ulfills the space tr ansportation needs of the United States.

Gentry, Jerauld R. "A Lifting Body Pilot Looks at Space Shuttle Requirements." *The Society of Experimental Test Pilots 1970 Report t o t he A erospace P rofession*. 10 (September 1970) : 179 -93. A p resentation a t th e 1 4th Symposium o f th is p rofessional o rganization h eld in B everly H ills, C alifornia, 2 4-26 S eptember 1970. I t discusses the r ole o f t he S huttle as a s pace v ehicle an d en dorses t he co ncept of a r eusable s ystem. I t al so suggests that jet engines are not necessary for the orbiter and that landing can be accomplished with glide only. This was a critical conclusion and one that was adopted by the program. This was based on the fact that lifting bodies had not been powered for landing and had worked fine.

Getting Aboard the Space Shuttle: Space Transportation System User Symposium. Piscataway, NJ: IEEE, 1978. This publication, the proceedings of a s ymposium, pr esents pa pers on v arious a spects of t he S huttle an d t he opportunities it provides. It is especially h elpful in a scertaining th e p ositions o n th e p rogram f rom th e standpoint of different users.

Gore, Rick. "When the Space Shuttle Finally Flies." *National Geographic*. 159 (March 1981): 317-47. In an article containing an ab undance o f t his p ublication's t rademark p hotographs, G ore o ffers an assessment of the development of the Shuttle through its first mission.

Guilmartin, John F., Jr., and Mauer, John Walker. *A Shuttle Chronology, 1964-1973*. Houston, TX: Lyndon B. Johnson Space Center, 1988. 5 Volumes. This is a co mprehensive and detailed chronology of the development of the Shuttle, divided into s eparate s ections co ncerning v arious as pects o f t he p rogram an d o rganized chronologically within them. It suffers from some repetition, but still has much valuable information and many reference notes.

Hanaway, John F., and Moorehead, Robert W. *Space Shuttle Avionics System*. Washington DC: National Aeronautics and Space Administration, 1989. This monograph describes the avionics systems of the Shuttle, celebrating the numerous "firsts" in the program: the incorporation of a co mprehensive fail operational/fail safe concept; the complex redundancy management techniques which became a s tandard in the industry; the use of digital data bus technology; the employment of high-order language to develop onboard software; the use of flight software program overlays from a tape memory; integration of flight c ontrol f unctions w ith th e r est o f the a vionics program; use of digital fly-by-wire technology; use of malfunction cathode-ray-tube display and crew interface approach; and the application of extensive operational services to onboard avionics systems.

Helms, W.R. "History, D esign, an d P erformance o f t he Space Shuttle H azardous G as D etection S ystem." *JANNAF Safety and Environmental Protection Subcommittee Meeting*. Lompoc, CA: NASA, 1984. pp. 195-202. This is a technical paper on the development of a critical safety feature of the Shuttle, prepared for presentation at this professional meeting.

Jeffs, George W. "The Space Shuttle: Its Interdisciplinary Design and Construction." *Interdisciplinary Science Reviews*. 4 (September 1979): 208-38. This is a lengthy scholarly article which surveys the scientific and engineering disciplines involved in the design and construction of the Shuttle to trace its evolution. It presents background material on the search for a reusable spacecraft and describes Shuttle operations and capabilities. It goes into detail to review the development of some Shuttle systems, especially as many technological areas were integrated. Jeff specifically looks at aerodynamics, propulsion, structural design, data processing and software, simulation exercises, crew training, verification testing and mission control.

Johnson, Colonel Roger W. "Advanced Space Programs: Transition to the Space Shuttle." *Astronautics & Aeronautics*. 14 (September 1976): 32-39. This is an intriguing discussion of the movement from the space program of the 1960s to the Shuttle as well as the movement of launches of such items as satellites to deployment by the Shuttle.

Kah, Carl L.C. *High Chamber Pressure Reusable Rocket Engine Technology*. New York: Society of Automotive Engineers, 1970. This technical work deals with the development of technology necessary to power the Shuttle. It is a good early statement of the status of the effort and the prospects for the future.

Kanai, K. "Applications of Active Control Technology to Spacecraft and Aircraft." *Journal of the Society of Instrument and Control Engineering*. 23 (January 1984): 157-62. Considers fly-by-wire and active control techniques and applications to the Space Shuttle.

Kranzel, Harold. "Shuttle Main Engine Story." *Spaceflight*. 30 (October 1988): 378-80. Although overshadowed after the *Challenger* accident and the attention focused on the solid-fuel boosters, the main engine has had a checkered history as well. The development, test, and problems of the main engine are noted in this short article, along with a table of all main engine flight events whether on an actual mission or a test.

Lewis, Richard. "Whatever Happened to the Space Shuttle?" *New Scientist*. 87 (31 July 1980): 356-59. Describes the reasons for the delays in the development of the Shuttle.

Loftus, J.P., Jr., *et al*. "The Evolution of the Space Shuttle Design." Unpublished paper written at the Johnson Space Center, Houston, TX, 1986. This report was prepared in response to requests by the Rogers Commission investigating the *Challenger* accident. It is a good technical discussion of the Shuttle's development. A copy is available at the History Office, Johnson Space Center.

Lore, Eugene S. "Advanced Technology and the Space Shuttle." *AIAA Paper 73-31*. Washington, DC: American Institute of Aeronautics and Astronautics, 1973. This was the tenth von Karman Lecture at the American Institute of Aeronautics and Astronautics held in Washington, DC, 8-10 January 1973. It presents a detailed account of NASA's overall technical challenge in the development of the Shuttle.

Lynch, Robert A. "The Space Shuttle Booster." Unpublished paper presented at the eighth Space Congress, Cocoa Beach, FL, 19-23 April 1971. This is an analysis and status report of the Shuttle booster configuration, presenting design features and performance characteristics. It analyzes the reasons behind choosing the delta wing over the stowed, fixed straight, or swept wing configurations, commenting that the delta wing was chosen on the basis of its compatibility to air breathing engine installation requirements rather than purely aerodynamic considerations.

Lyndon B. Johnson Space Center. *Technology Influence on the Space Shuttle Development*. Houston, TX: Johnson Space Center, 1986. Because of desired low development costs, designers of the Space Shuttle used existing technology whenever possible. This increased maintenance costs and turnaround time to the point that the Shuttle has been unable to obtain the expected low refurbishment and reuse rates. The report concludes that technological emphasis should be placed on maintainability, refurbishment, and reuse.

Malkin, M.S. "Space Shuttle/The New Baseline." *Astronautics & Aeronautics*. 12 (January 1974): 62-78. Written by the director of the Space Shuttle program office at NASA, this lengthy article is a detailed rundown of the development of the Shuttle through 1973. It contains a wealth of information about the Shuttle's characteristics, dimensions, and capabilities. Virtually every important system in the Shuttle is described in some way, and a host of illustrations help with this process. There is also a description of the proposed flight of the Shuttle from both Cape Canaveral and Vandenberg Air Force Base as well as analyses of trajectories and recovery data. It concludes with a statement of faith that the Shuttle will be built on a reduced budget but with essentially the same capabilities as originally proposed.

McKenzie, P.J. "Structural Review of the Space Shuttle." *Journal of the British Interplanetary Society*. 26 (October 1973): 597-605. Surveys the preliminary research on reusable spacecraft from which the Shuttle emerged, and asserts that the modified system will not support the development of sophisticated primary structures and metallic thermal protection systems.

Michel, Rudi. "Propulsion Systems for Single-Stage Shuttles." *Astronautics & Aeronautics*. 2 (November 1974): 32-39. Describes and analyzes the plans for the launch vehicles supporting the Shuttle orbiter.

Morea, S.F., and Wu, S.T. eds. *Advanced High Pressure Oxygen/Hydrogen Technology*. Washington, DC: National Aeronautics and Space Administration, 1985. This publication contains the proceedings of a conference on rocket propulsion held at the Marshall Space Flight Center, Huntsville, AL, on 27-29 June 1984.

Mueller, George E. Address on the Space Shuttle before the British Interplanetary Society, University College, London, England. August 10, 1968. Copy in National Aeronautics and Space Administration Reference Collection, NASA History Office, Washington, DC. This presentation, made by NASA's Associate Administrator for Manned Space Flight, may well have been the first public presentation of the Shuttle concept to a scholarly community. It set up the rationale, technological choices, and planning activities taking place at NASA for the development of the Space Transportation System.

NASA's Plans to Procure New Shuttle Rocket Motors. Washington, DC: U.S. House of Representatives Committee on Government Operations, 1986. This lengthy report contains the hearing of the Legislation and National Security Subcommittee on this subject conducted on 31 July 1986 after the loss of *Challenger*.

NASA Space Shuttle Technology Conference. Washington, DC: National Aeronautics and Space Administration, 1971. NASA Technical Memorandum (TM) X-2272, X-2273, and X-2274. This three volume work contains the presentations of a conference held at Langley Research Center, Hampton, Virginia, 2-4 March 1971. Each volume is dedicated to a specific area of consideration: I-Aerothermodynamics, Configurations, and Flight Mechanics; II-Structure and Materials; and III-Dynamics and Aeroelasticity.

"Orbiter Protective Tiles Assume Structural Role." *Aviation Week & Space Technology*. 25 February 1980, pp. 22-24. Although a news story, this article is an excellent report on the development and use of the special tiles on the orbiter used to absorb heat during reentry.

Poll, Henry O., et al. *Space Shuttle Reaction Control System*. New York: Society of Automotive Engineers, 1970. This is a short technical publication dealing with the development of one aspect of the Shuttle's control system.

Rainey, Robert W. "Progress and Technology for Space Shuttles." Unpublished paper presented at the sixteenth annual meeting of the American Astronautical Society, Anaheim, CA, 8-10 June 1970. *AAS Paper 70-046*. This is a fine technical paper on the efforts to develop a reusable spacecraft written by a senior engineer at the Langley Research Center. Rainey notes that during the past several years, considerable effort has been expended by industry and government to define low-cost transportation systems envisioned to operate from earth to orbit and return. The most recent NASA studies, the Phase A Integral Launch and Reentry Vehicle (ILRV) Studies, were completed in the latter part of 1969. The study's goals were to determine the feasibility of Shuttle vehicles that would reduce the cost per pound of payload to orbit by an order of magnitude with improvements in systems reliability and crew safety. Primary emphasis was place upon two-stage fully reusable systems, and in this paper, designs of the Phase A Shuttles are examined. Special attention is given to contractual and in-house activities in several key technological areas: structures, thermal protection, aerothermodynamics, aerodynamics, and approach and landing.

Robinson, John W. ed. *Shuttle Propulsion Systems*. New York: American Society of Mechanical Engineers, 1983. This book collects presentations made at the annual meeting of the American Society of Mechanical Engineers on 14-19 November 1982 relating to the development and technology of the Shuttle propulsion system.

Salkeld, Robert, and Skulsky, R.S. "Air Launch for Space Shuttles." *Acta Astronautica*. 2 (July/August 1975): 703-713. Explores some of the possibilities for launching the Shuttle not vertically on a booster like ordinary rocket payloads but from aircraft for flight into the upper atmosphere and eventually to space using its aerodynamic features.

Salkeld, Robert. "Single-Stage Shuttle for Ground Launch and Air Launch." *Astronautics & Aeronautics*. 17 (June 1979): 52-64. This article describes the technical effort going into the study of launch vehicles for the Shuttle. Salkeld stresses the diversity of design potential that decision makers should grasp to shape effective space transportation. Various and diverse types of single-stage Shuttles are identified in terms of mission and operational capabilities, physical characteristics and economics. It is shown that the development of economical space transportation can be realistically regarded as feasible, and that it will make possible the performance of dependable, commercially viable operations beyond the earth. With this system it should be possible to bring such systems into operation within 10-15 years. The development of single-stage Shuttles is expected to represent a significant advance in space travel.

Salkeld, Robert. "Space Shuttle: Some Growth Possibilities." *Spaceflight*. 15 (November 1973): 402-408. This article looks at the Shuttle as a means to develop new capabilities and exploit new technologies. Argues that the single-stage concept of the Shuttle appears particularly interesting in view of the possibility that such a vehicle could be converted directly to a global-orbiter commercial transport by exchanging rockets for jets. Salkeld considers other growth options: replacing the solid rocket motors with a fully reusable booster, replacement of both solid motors and the main tank with reusable boosters, and replacement of the entire baseline Shuttle with a fully reusable single-stage vehicle. For almost all growth concepts, irrespective of vehicle configuration, mixed-mode propulsion promises significant advantages in the form of improved performance, reduced vehicle size and manufactured hardware weight, and reduced propellant costs.

Scherer, Lee R. "Space Transportation System: Status Report." *British Interplanetary Society Journal*. 32 (October 1979): 364-70. A useful synopsis of the development of the Shuttle through its flight tests, written for a British audience. The emphasis is on hardware and the overall system.

Scott, Harry A. "Space Shuttle: A Case Study in Design." *Astronautics & Aeronautics*. 17 (June 1979): 54-58. This is a brief review of the evolution of the Shuttle design describing how one contractor involved in the studies, Rockwell International, worked through the process. It emphasizes the changing requirements for the Shuttle, as well as the difficult economic problems the program faced, during the course of these studies. It has a good degree of technical information and discusses how the DOD requirements for cross-range capability prompted the change from a straight to delta wing configuration. Scott's conclusion that "the orbiter's performance requirements were reasonably maintained and achieved even though the program's projected funds were halved" is debatable.

Sheridan, Caroline T. "Space Shuttle Software." *Datamation*. 24 (July 1978): 128-40. A useful article on the problems, priorities, and promise of the computer systems being developed for use in the Shuttle.

The Space Shuttle—Its Current Status and Future Impact. Warrendale, PA: Society of Automotive Engineers, 1981. A collection of 12 technical presentations on the Shuttle and various aspects of the program. Maxime Faget offers perhaps the most useful article for the nonspecialist on the overall development of the Shuttle.

Space Shuttle Main Engine Development Program. Washington, DC: U.S. Senate Commerce, Science, and Transportation Committee, 1978. This is the report of a hearing on the Shuttle's engine development difficulties before the 95th Cong., 2d Sess.

Space Shuttle Missions of the '80s. San Diego, CA: American Astronautical Society, 1976. This book is volume 32 of the society's "Advances in Astronautical Sciences" series. It speculates on the nature and extent of Shuttle missions of the future and suggests that routine access to space will come as a result of these missions, space science will blossom, and the exploration of the planets will be possible. It is a promotional work, but based on realistic possibilities.

Space Shuttle Program: Proceedings of the Short Course, Boulder, CO, October 6-7, 1972. Boulder, CA: American Institute of Aeronautics and Astronautics, 1973. The Space Shuttle program is surveyed in several papers dealing with project management and planning functions, design concept definition studies, projected mission profiles, vehicle hardware and systems configurations, testing programs, and mission support requirements. Specific topics considered include typical payload capabilities in various types of missions, flight operations concepts, aerodynamic aspects of the orbiter, thermal protection systems, design and performance of the main engines, design of the external drop tank and solid rocket motor stages, recovery and refurbishment of the Shuttle, and launch center tasks and facilities.

Space Transportation System User Handbook. Washington, DC: National Aeronautics and Space Administration, June 1977, rev. ed. May 1982. This loose-leaf booklet explains the Shuttle's development and uses. It describes the types of user support it was designed for and offers some pricing background.

Strouhal, George, and Tillian, Donald J. "Testing the Shuttle Heat Protection Armor." *Astronautics & Aeronautics.* 14 (January 1976): 57-65. This is a technical paper on the design, construction, and testing of heat protection systems on the Shuttle. Specific discussion of the heat-absorbing tiles on the outside of the orbiter is included.

Talay, T.A.; Morris, W.D.; Eide, E.G.; and Rehder, R.R. "Designing for a New Era of Launch Vehicle Operational Efficiency." *Astronautics & Aeronautics.* 21 (June 1983): 44-48. The authors contend that now is the time to start work on a next-generation Shuttle, and that early and explicit consideration of operational requirements and assessment of their effects provide the best means of designing an economically viable system. Some of the operational features affecting configuration design are fleet size, operational mode, refurbishment, and resource requirements. The proposed system has a simplified operational role in the Shuttle, which, in addition to transportation, also had to allow experiments, support payloads, and stay longer than a week in orbit. Once the Space Station is in orbit, it will only be required to provide transportation. The authors advocate the development of a two-stage, fully-reusable launch vehicle designed to carry 150,000 pounds to a space station, off-load, and return. It would always be launched fully loaded and its cargo would be processed and redistributed at the space station. The authors give some attention to ground servicing, flight operations, rendezvous-compatible orbits, launch windows, standard trajectories, entry windows, operational costs, the mission model, and resource requirements.

Tischler, A.O. "A Commentary on Low-Cost Space Transportation." *Astronautics & Aeronautics.* 7 (August 1969): 50-64. A well-done technical discussion of the ability of the United States to develop a reusable, and therefore low-cost, space vehicle as the follow-on system for the Saturn rocket. Written by the head of the NASA Office of Advanced Research and Technology, it suggests that "the pace of payload development, the availability of manpower and funds, and the technological state of the art argue a transitional approach to fully recoverable large space transports" (p. 50).

Tischler, A.O. "Defining a Giant Step in Space Transportation: Space Shuttle." *Astronautics & Aeronautics*. 9 (February 1971): 22-67. This is a special section in this publication relating to the Shuttle's development as it stood early in the program. Tischler had specialists write on the following: technology for aerothermodynamics, structures, dynamics and aeroelasticity, life support, protective systems, crew system interface, and propulsion. It represents a very good technical overview of the major fronts on which the Shuttle was moving in 1970-1971.

Tischler, A.O., ed. *Space Transportation System Technology Symposium*. Cleveland, OH: Lewis Research Center, 1970. Technical Memorandum (TM) X-52876. This seven-volume work is the published proceedings of a symposium on the Shuttle held at Lewis Research Center on 15-17 July 1970. Consisting of papers on various subjects, the volumes are: I-Aerothermodynamics and Configurations; II-Dynamics and Aeroelasticity; III-Structures and Materials; IV-Propulsion; V-Operations, Maintenance, and Safety; VI-Integrated Electronics; and VII-Biotechnology.

Townsend, Marjorie R. "Direct Delivery of Automated Spacecraft Using the Shuttle: Thoughts for the Designer." *Astronautics & Aeronautics*. 15 (April 1977): 32-38. Argues that the Shuttle should be employed for the movement of special items between earth and orbit for future exploration beyond the planet. It offers some thoughts on how the Shuttle should be built to make this more practicable.

Vaughn, Robert L. ed. *Space Shuttle: A Triumph in Manufacturing*. Dearborn, MI: Society of Manufacturing Engineers, 1985. This collection of essays, written by members of the society of Manufacturing Engineers, discusses the construction of the Shuttle.

Von Braun, Wernher. "The Reusable Space Transport." *American Scientist*. 60 (November-December 1972): 730-38. This is a lucid description of the Shuttle as envisioned near the time of its approval by President Nixon. It describes the components of the Shuttle's configuration and in the best sense of the term tries to sell the program, arguing that the system will be versatile and the keystone of the continuing exploration and use of space.

Weidner, Don K., ed. *Space Environment Criterial Guidelines for Use in Space Vehicle Development (1969 Revision)*. Huntsville, AL: Marshall Space Flight Center, 1969. This document provides a cross-section of national environmental data for use as design criteria guidelines in the space vehicle development program. The document focuses on earth science issues, especially atmospheric considerations.

Whitsett, C.E., Jr. "Manned Maneuvering Units." *Space Shuttle and Spacelab Utilization: New-term and Long-Term Benefits for Mankind*. San Diego, CA: Univelt, 1978. pp. 617-31. This article, delivered at the sixteenth Goddard Memorial Symposium in 1978, focuses on the manned maneuvering units that the Shuttle crews would use in orbit. The new space suit with attached life-support system and self-contained propulsion backpack would allow the crew to venture beyond the confines of the cargo bay. It would allow free-flight from the Shuttle cargo bay to satellites and return. The freedom of movement this allows would make working in space a realistic capability. The paper describes the suit and its method of operation.

Yaffee, Michael L. "Alternate Booster Evaluation Set." *Aviation Week & Space Technology*. 24 January 1972. pp. 36-37. This article reports on the efforts of Grumman and Boeing in investigating the use of a pump-fed booster as an alternative to the pressure-fed ballistic recoverable booster that had been intended for the Shuttle.

Link to Part 2 (1992–2011), Chapter 4—Design and Development

CHAPTER 5
SPACE SHUTTLE TESTING AND EVALUATION

Baker, David. "Evolution of the Space Shuttle: 5. Approach and Landing Test Programme." *Spaceflight*. 19 (June 1977): 213-17. As the subtitle suggests, this article describes the major developments in the test program for the Shuttle's reentry and landing phase.

Baker, David. "Evolution of the Space Shuttle: 6. Free Flight Tests Begin." *Spaceflight*. 20 (January 1978): 21-28, 40. Describes the flight characteristics of the Shuttle tests then underway at the Dryden Flight Research Center in California.

Baker, David. "Space Shuttle Feels the Atmosphere." *Flight International*. 26 March 1977, pp. 784-787. Summarizes the flight tests of the first Shuttle orbiter then being completed at the Dryden Flight Research Center.

Baker, David, and Wilson, Michael. "Space Shuttle Debut." *Flight International*. 25 September 1976, pp. 975-982. Description of the first roll-out of the Shuttle *Enterprise* in 1976.

Beatty, J. Kelly. "Space Shuttle: Problems and Progress." *Sky and Telescope*. 57 (June 1979): 542-44. Discusses the Shuttle's construction and testing as of 1979. Its premise is whatever is good for the Shuttle is good for astronomers.

Beatty, J. Kelly. "The Space Shuttle in Free Flight." *Sky and Telescope*. 54 (October 1977): 271-275. News report on the flight tests of the Shuttle after being deployed from the Boeing 747 used in its flight testing.

Bulloch, Chris. "Space Shuttle Progress." *Interavia*. 35 (October 1980): 899-906. This article describes for an international audience the development of the Shuttle from its conception through its flight tests.

Covault, Craig. "Shuttle Aborts Pose New Challenges." *Aviation Week & Space Technology*. 15 October 1975, pp. 39-45. Reports in depth on the difficulties of the Shuttle after one of its tests were aborted due to malfunctions.

Covault, Craig. "Shuttle Engine Passes Critical Milestone." *Aviation Week & Space Technology*. 30 June 1975, pp. 37-42. This is one of many news articles on the Shuttle by Covault. Discusses the major testing of the Shuttle main engine.

Covault, Craig. "Shuttle Firing Test Count Starts." *Aviation Week & Space Technology*. 9 February 1981, pp. 24-26. An account of the test firing of the Shuttle's main engines.

Edwards, C.L.W., and Cole, Stanley R. *Predictions of Entry Heating for Lower Service of Shuttle Orbiter*. Washington, DC: National Aeronautics and Space Administration, 1983. This 94-page report is a technical memorandum on this subject.

Fink, Donald. "Orbiter Experiences Control Problems." *Aviation Week & Space Technology*. 31 October 1977, pp. 16-17. Describes the problems found in the handling of the Shuttle during its free flight tests. On its fifth free flight test on 26 October 1977 the *Enterprise* encountered control problems at touchdown. While trying to slow the spacecraft for landing the pilot experienced a left roll, corrected for it, and touched down too hard. The Shuttle bounced once and eventually settled down to a longer landing than was expected.

Fink, Donald. "Orbiter Flight Plan Expanded." *Aviation Week & Space Technology*. 27 June 1977, pp. 12-14. This news article describes the current status and fast pace of the Shuttle test program at Dryden in the latter 1970s.

Fink, Donald. "Orbiter Responsive in Free Flight." *Aviation Week & Space Technology*. 22 August 1977, pp. 12-19. This is a news report on the tests of the Shuttle orbiter from the Boeing 747 test bed.

Flight Test Results Pertaining to the Space Shut tlecraft. Washington, DC: National Aeronautics and Space Administration, 1970. Available from Springfield, VA: Federal Scientific and Technical Information. This 159-page work was published as NASA Technical Memorandum (TM) X-2101. It contained the proceedings of a symposium held at the Flight Research Center at Edward Air Force Base, California, on 30 June 1970, assessing the status of the lifting body program and various aspects of the Shuttle program then being developed. It is a highly technical set of presentations.

Gong, Leslie; Ko, William L.; and Quinn, Robert D. *Thermal Response of Space Shuttle Wing During Reentry Heating*. Edwards, CA: Dryden Flight Research Facility, 1984. This is a highly technical report concerning the reentry problems of the Shuttle orbiter.

Grey, Jerry. *Enterprise*. New York: William Morrow and Co., 1979. This is a popularly-written book on the decision, development, and test of the early Shuttle, the Orbiter 101, named after the "Star Trek's" *Enterprise*. Designed for the popular market, it is a fast-moving story with emphasis on anecdotes, and without scholarly apparatus.

Hallion, Richard P. *Test Pilots: The Frontiersmen of Flight*. Washington, DC: Smithsonian Institution Press, 1988 ed., pp. 270-80. This addendum to the 1981 original edition of this book reviews the test and evaluation phase of the Space Shuttle. It is a good summation of the development of the program through the first flights of the orbiter.

Hamilton, David D.; Schliesing, John A.; and Zupp, George A., Jr. *Orbiter Loads Math Model Description and Correlation with ALT Flight Data*. Springfield, VA: National Technical Information Service, 1980. This is a short technical paper relating to flight operations and tracking of the Shuttle.

Kolcum, Edward H. "Shuttle Engine Firing Successful." *Aviation Week & Space Technology*. 2 March 1981, pp. 17-19. Discussion of the successful test of the Shuttle's main engines.

Lenorovitz, Jeffrey M. "Shuttle Orbiter Test Phase Trimmed." *Aviation Week & Space Technology*. 4 July 1977, pp. 18-19. This news item describes the conduct of the captive test flights of the Shuttle mated to the Boeing 747 at Dryden and comments that since they were going so well the test program was ahead of schedule and the number of flights could be cut because they were unnecessary.

McIntosh, Gregory P., and Larkin, Thomas P. "Space Shuttle's Testing Gauntlet." *Astronautics & Aeronautics*. 14 (January 1976): 44-56. Technical article on the rigid testing program given for each system of the Shuttle. It emphasizes the safety aspects of the program.

National Research Council. *Technical Status of the Space Shuttle Main Engine (Second Review)*. Washington, DC: National Research Council, February 1979. This is the second of two reports of the Assembly of Engineering's ad hoc Committee for Review of the Space Shuttle Main Engine Development Program. A follow-up study was requested by the Subcommittee on Science, Technology, and Space of the U.S. Senate Committee on Commerce, Science, and Transportation as the result of uncertainties in technical development pointed out in the committee's first report. This report presents the committee's assessment of problems considered in its earlier review as well as others that arose subsequently. It also addresses longer-range issues concerning safety and reliability. Damaging malfunctions in two main engines while under test during the period of the committee's study are described together with a discussion of their implications for development of the main engines.

Whitnah, A.M., and Hillje, E.R. *Space Shuttle Wind Tunnel Testing Program Summary*. Washington, DC: National

Aeronautics and Space A dministration, 1984. Detailed en gineering r eport d escribing te sts o f th e S huttle in wind tunnels.

Link to Part 2 (1992–2011), Chapter 4—Design and Development

CHAPTER 6
SPACE SHUTTLE OPERATIONS

Arrington, James P., and Jones, Jim J. Comps. *Shuttle Performance: Lessons Learned*. Washington, DC: National Aeronautics and Space Administration, 1983. 2 Vols. This is a collection of papers given at a conference on the Shuttle held at Langley Research Center on 8-10 March 1983 for the purpose of ascertaining the operational performance of the Shuttle after its first missions into orbit. It has several papers dealing with a number of broad areas: ascent aerodynamics; entry aerodynamics; guidance, navigation, and control; aerothermal environment; thermal protection; and measurements and analyses.

Carrillo, Manuel J. *A Development of Logistics Management Models for the Space Transportation System*. Santa Monica, CA: Rand Corp., 1983. This study reviews procedures and sets priorities and policies for the support of Shuttle operations.

Case, Ed. *"We Have Lift Off": History and Photos of Shuttle Launches*. Palm Bay, FL: E. Case, 1989. Collection of photos of Shuttle launches, some of them striking, with a little text about the launches.

Covault, Craig. "Aerobatics at Mach 25." *Science 81*. 2 (May 1981): 58-63. A short article on the first Shuttle mission in 1981.

Covault, Craig. "Planners Set Long-Term Space Goals." *Aviation Week & Space Technology*. 9, March, 1981, pp. 75-78. This article reports on NASA and other Federal Government leaders' efforts to assess the direction of the space program for the future as the Shuttle begins its operational phase.

Covault, Craig. "Shuttle Management Shifting to Operations." *Aviation Week & Space Technology*. 21 December 1981, pp. 12-15. This article reviews the process of change taking place in NASA as it moved from an RDT&E stance with the Shuttle to one oriented toward flying operational missions.

Dawson, Harry S. *Review of Space Shuttle Requirements, Operations, and Future Plans*. Washington, DC: U.S. House of Representatives Committee on Science and Technology, 1984. This report deals with the past and prospects for the Shuttle during its early operational life. It is optimistic but still not sanguine that NASA would be able to make it cost effective. This report was prepared by the House Subcommittee on Space Science and Applications.

General Accounting Office. *Space Program: Space Debris a Potential Threat to Space Station and Shuttle*. Washington, DC: General Accounting Office, 1990. This report, written for the chair of U.S. House of Representatives Committee on Science, Space, and Technology, deals with the problem of orbiting junk and its potential hazard to the Shuttle and other flight operations in the next five years.

Gore, Rick. "When the Space Shuttle Finally Flies." *National Geographic*. 159 (March 1981): 317-47. In an article containing an abundance of this publication's trademark photographs, Gore offers an assessment of the development of the Shuttle through its first mission.

Gurney, Gene, and Forte, Jeff. *The Space Shuttle Log: The First 25 Flights*. Blue Ridge Summit, PA: Aero Books, 1988. Briefly covers each of the first 25 flights of the Shuttle in chronological order. Each mission begins with a flight detail entry covering the basics of the missions, personnel, payload, experiments, etc. Nothing in this book is really new.

Kolcum, Edward H. "Managers Modernize Shuttle System to Increase Efficiency, Launch Rate." *Aviation Week & Space Technology*. 4 December 1989, pp. 46-48. This article focuses on the efforts of Robert L. Crippen, astronaut and manager of the Shuttle program, to reorganize his function to ensure safe and reliable operations.

Lewis, Richard S. *The Voyages of* **Columbia**: *The First True Spaceship*. New York: Columbia University Press, 1984. Taking as its theme that the Shuttle is the first true spaceship—one that can be reused and makes access to space more routine—this book provides a good rendition of the development and use of the *Columbia* orbiter. A large-format, well-written publication, it has numerous photographs and illustrations, as well as scholarly notes. There is much discussion of development and testing, procedures for operations such as solid-rocket booster use and recovery, and a detailed account of each mission. Probably the best book of its type, its focus and theme is limited to a single orbiter and its role in the space program.

National Aeronautics and Space Administration. *Space Transportation System User Handbook*. Washington, DC: National Aeronautics and Space Administration, June 1977, rev. ed. May 1982. This loose-leaf booklet explains the Shuttle's development and uses. It describes the types of user support it was designed for and offers some pricing background.

Oberg, Alcestis R. "After the Parades." *Final Frontier*. September/October 1990, pp. 43-48, 58-59. This article describes the little-known or little-noticed efforts to refurbish the Shuttle orbiters after each flight. After flight they are in terrible condition and are sent to a Shuttle garage for a complete overhaul. The author details the work done on the various systems, etc., to get the spacecraft ready to fly again.

Overbye, Dennis. "The Shuttle Comes of Age." *Discover*. June 1982, pp. 61-64. Published just after the first few operational flights of the Shuttle, this short article assesses for a popular audience the importance of the craft for the United States.

Phillips, W. Pelham. *Space Shuttle Orbiter Trimmed Center-of-Gravity Extension Study*. Hampton, VA: Langley Research Center, 1984. This is one of several technical studies undertaken during this period to correct for flight characteristics of the Shuttle.

Powell, Joel W. and Caldwell, Lee Robert. *The Space Shuttle Almanac: A Comprehensive Overview of the First Ten Years of Space Shuttle Operations*. Calgary, Alberta: Microgravity Press, 1992. This reference tool offers an overview of Space Shuttle operations, facilities, hardware, and missions for the Shuttle's first 39 flights. Using information culled mainly from various NASA publications, the authors describe the orbiter, discuss each mission flown, and provide detail on payloads, experiments, and crew.

Results of Space Shuttle Flight 41-C. Washington, DC: U.S. House of Representatives Committee on Science and Technology, 1984. This contains hearings about this Shuttle flight held on 15 May 1984.

Results of Space Shuttle Flight 41-D. Washington, DC: U.S. House of Representatives Committee on Science and Technology, 1984. This contains hearings about this Shuttle flight held on 25 September 1984.

Results of Space Shuttle Flight 61-C. Washington, DC: U.S. House of Representatives Committee on Science and Technology, 1986. This contains hearings about this Shuttle flight held on 23 September 1986.

Results of the Space Shuttle Discovery Mission. Washington, DC: U.S. Senate Committee on Commerce, Science, and Transportation, 1988. This publication reports on the hearings of the Subcommittee on Science, Technology, and Space held on 13 October 1988.

Space Shuttle Payloads: Hearing Before the Committee on Aeronautical and Space Sciences. Washington, DC: U.S. Senate Committee on Aeronautical and Space Sciences, 30-31 October 1973. This two-part publication reviews the Shuttle's possible missions in the 1980s.

Space Shuttle Requirements, Operations, and Future Plans. Washington, D C: U .S. H ouse o f R epresentatives Committee o n S cience an d T echnology, 1 984. T his co nsists o f t he h earings o n the subject before the subcommittee on Space Science and Applications of the 98th Cong., 2d Sess.

Space Shuttle Transportation System: Press Information. Downey, C A: R ockwell I nternational, F ebruary 1981. T his publication contains facts and information about the Shuttle for the media. It has been issued periodically since the system first began operation, updating certain aspects of the program.

Stockton, William, and W ilford, J ohn N oble. *Spaceliner: R eport on C olumbia's V oyage i nto T omorrow.* Ne w Yo rk: Times B ooks, 1 981. T his is a p opular d iscussion o f th e d evelopment a nd f light o f th e f irst S huttle m ission, *Columbia*, in 1981. It is heavy on fast-paced narrative and anecdotes, and thin on documentation. It keeps the human element of the story in the forefront, and while there is some discussion of technological developments, those are certainly subservient to the good story the authors try to tell.

Trippett, Frank. "Milk Run to the Heavens." *Time.* 12 January 1981, pp. 10ff. A well-done news story with characteristic striking photography, this article describes the use of the Shuttle for routine operations. No longer would space flight be a difficult venture, but one that is, as the writer said, a "milk run."

Link to Part 2 (1992–2011), Chapter 5—Operations

CHAPTER 7
CHALLENGER ACCIDENT AND AFTERMATH

Assured Access to Space: 1986. Washington, DC: U.S. House of Representatives Committee on Science and Technology, 1986. This publication reports on hearings held by the Subcommittee on Space Science and Applications on several occasions in 1986 following the *Challenger* accident.

Baker, David. "Science Crashed with Challenger." *New Scientist*. 29 January 1987, pp. 55-57. Written on the first anniversary of the *Challenger* accident, this article asserts that space science was set back many years because of the retrenchment of the space program. Baker analyzes what he thought was a dangerous trend in the space program, the subversion of science for military payloads. Even without this, contends Baker, literally hundreds of scheduled experiments for the Shuttle have been delayed for an indefinite period. He ends by saying that the fate of the Shuttle resulted in the loss of 38 years from the science projects discussed in the article.

Beck, Melinda. "NASA's Troubled Flight Plan: There's no Turning Back, but Are We on the Right Path?" *Newsweek*. 10 February 1986, pp. 35-38. This is a good article on the development of the space program and the alleged loss of nerve resulting from the explosion of *Challenger*. It suggests that NASA was naive to put all eggs in the Shuttle basket and that a reevaluation is appropriate.

Bell, Trudy E. and Esch, Karl. "The Fatal Flaw in Flight 51-L." *IEEE Spectrum*. (February 1987): 36-51. This article, based on personal interviews and Rogers Commission and congressional committee reports, reviews the events leading up to the *Challenger* accident. It discusses the steady erosion of concern about the deficiencies of the O-ring design as NASA and Thiokol became more complacent with the Shuttle's operational success. Other reasons cited for NASA's and Thiokol's failure was their rigid chains of command, the lack of input from the NASA safety office, the financial and possible political pressure to launch, and the lack of secondary manufacturing sources for the boosters. The article concludes that management was in general too wedded to a climate that simply ignored bad news, rather than pay attention to it and try to correct reported problems.

Bell, Trudy E., and Esch, Karl. "The Space Shuttle: A Case of Subjective Engineering." *IEEE Spectrum*. (June 1989): 42-46. These writers, in an unreferenced article based on interviews, describe the development of a NASA cultural bias toward engineering reliability and safety during the development and construction phase of a spacecraft. Because of this approach, NASA never placed reliance on probabilistic risk analysis, the statistical tracking of failure rates, and had collected none on which to base a statistical analysis. This was not an issue until the *Challenger* accident, when it was found that a statistical effort could have predicted the probability of failure and perhaps signaled that caution was in order.

Biddle, Wayne. "NASA: What's Needed to Put it on its Feet?" *Discover*. 8 (January 1987): 30-41. This is a lengthy special report on issues relating to NASA management of the space program. The Shuttle program, its management, operations, and restructuring following *Challenger* are examined.

Brosz, Tom. "NASA Works to Get Shuttle Back into Space." *Commercial Space Report*. 10 (November 1986): 1-6. This article discusses NASA's efforts to recover its Space Shuttle program following the *Challenger* tragedy and to begin flight once again. Brosz suggests that the majority of the payloads are military and government satellites. NASA hopes to restructure the Shuttle program's management system and develop a management structure based upon the successful Apollo program. It was also working to correct flaws in the solid rocket boosters and crew escape systems.

Brosz, Tom. "The Challenger Disaster: Causes and Consequences—Part II." *Commercial Space Report*. 10 (March 1986): 1-5. According to the author, NASA's monopoly on all Space Shuttle flights was the cause of what became an essential prohibition from space for the United States following the *Challenger* accident. Because of the monopoly, commercialization of the Space Shuttle had been virtually impossible, and now that the accident has occurred, companies are forced to look for alternative launch sites and vehicles. Europe's Arianespace may try to absorb as many of the satellite customers as it can. With all the space eggs in the Shuttle basket, the United States has been forced to take a back seat to other nations who have a variety of launch capabilities.

Challengers: The Inspiring Life Stories of the Seven Brave Astronauts of Shuttle Mission 51-L. New York: Pocket Books, 1986. This book, written by the staff of the Washington *Post*, describes the careers of the seven astronauts killed in the *Challenger* accident of January 1986. It is a relatively standard journalistic account, but it contains considerable background not found elsewhere.

Cohen, Daniel and Susan. *Heroes of the Challenger.* London: Archway Paperbacks, 1986. Published in July 1986 about the people killed in the Challenger explosion. Provides a media approach to covering the life stories and events of STS-51L in an inexpensive and compact version of *Challengers*.

Dobrzynski, Judith H. "Morton Thiokol: Reflections on the Shuttle Disaster." *Business Week*. 14 March 1988, pp. 82-83. Discusses the problems and accomplishments of Morton Thiokol in light of the O-ring problem on the solid rocket boosters that was a critical failure item for the *Challenger* accident. Special attention is paid to the work of Charles S. Losk, lead man at Morton Thiokol.

Dworetzky, Tom. "Return of the Shuttle." *Discover.* 7 (July 1988): 46-55. This is a special report somewhat flippantly subtitled "Righting the Stuff," which surveys the space program. A significant portion of it deals with the Shuttle program and how NASA is restructuring it to return to space following the *Challenger* accident.

Feynman, Richard P. "An Outsider's Inside View of the Challenger Inquiry." *Physics Today.* 41 (February 1988): 26-37. This article, by one of the nation's leading scientists, is an outstanding discussion of the O-ring problem and Thiokol's attempted solution both before and after the Shuttle accident.

Feynman, Richard P. *"What Do You Care what Other People Think?" Further Adventures of a Curious Character.* New York: W.W. Norton and Co., 1988, as told to Ralph Leighton. This is a fascinating personal account of the work of the Rogers Commission to unravel the reasons behind the *Challenger* disaster. Feynman was responsible for the assignment of responsibility for the accident to the O-rings of the solid rocket boosters and he discusses with compassion and insight the personalities and events surrounding the discovery.

Forres, George. *Space Shuttle: The Quest Continues.* London: Ian Allen, 1989. In what could only be considered a broad introduction to the Shuttle program, the author describes the reassessment of the Shuttle program following the *Challenger* accident and its return to flight in September 1988. Designed for the buff market, it is well-illustrated but has no references.

Furniss, Tim. "Space Comes Down to Earth." *Space.* 2 (September-November 1986): 38-41. This article examines the *Challenger* accident and other failures and how they may affect the future exploration of space. A brief overview of the Shuttle's budget, missions, setbacks, and schedules is given. Some of the pressures on the Shuttle that he identifies include a wide range of customers, the space station, Shuttle modifications, presidential reports, and safety. The author offers a provisional flight schedule for the Shuttle.

General Accounting Office. *Space Shuttle: Changes to the Solid Rocket Motor Contract TLSP: Report to Congressional Requestors.* Washington, DC: General Accounting Office, 1988. This report, done after the *Challenger* accident, describes the changes to the Space Shuttle solid rocket motor contract, and assesses the redesign of the motors following the accident, describing the changes in the motor joints and other design changes to enhance the motor's safety and reliability. These changes were incorporated into 13 sets of boosters for the Shuttle. It also comments on the method used to assess the costs of these changes, noting that the fees paid were changed from specific cost and performance incentives to more subjective valuations by NASA.

General A ccounting O ffice. *Space Shut tle: F ollow-up Evaluation of NASA's Sol id R ocket M otor P rocurement.* Washington, DC: General Accounting Office, 1989. T his r eport r eviews N ASA's " plans to e stablish a nd maintain c ompetition in the future procurement of Shuttle solid rocket motors and the quality assurance and industrial safety programs at Morton Thiokol's solid rocket motor manufacturing plant."

General A ccounting O ffice. *Space Shut tle: N ASA's P rocurement of Solid R ocket B ooster M otors.* Washington, DC: General A ccounting O ffice, A ugust 1 986. T his r eport as sesses N ASA's ef forts t o p rocure the solid rocket booster from Morton Thiokol following the *Challenger* accident. It discusses the redesign effort and analyzes the costs to the Shuttle program.

Goldberg, Steven. "The Space Shuttle Tragedy and the Ethics of Engineering." *Jurimetrics Journal.* 27 (Winter 1987): 155-159. T his ar ticle cr iticizes t he d ivorce o f en gineering judgment from management d ecisions, using the *Challenger* accident as a case study.

Implementation of t he R ecommendations of t he P residential C ommittee on t he Spac e Shut tle Challenger Accident. Washington, DC: National Aeronautics and Space Administration, June 1987. This is an in-depth report of the NASA leadership to President Ronald Reagan on the method and results of the implementation of changes recommended to th e S huttle p rogram in lig ht o f th e *Challenger* accident. The l eadership em phasized t he sweeping changes coming to the agency to ensure that nothing of this type ever happened again. They note that the interplay of national politics, NASA management, and individual engineering decisions is critical.

Investigation of the Challenger Accident: Hearings before the Committee on Science and Technology. Washington, DC: U.S. House of Representatives Committee on Science and Technology, 1986. 2 Vols. This large work contains the t estimonies o f num erous hi gh-ranking witnesses c oncerning t he *Challenger* accident. I t i ncludes t he statements of more than 60 people taken during 10 formal sessions. Understanding its contents is critical in any serious effort to appreciate and interpret the complexity of the events leading up to the tragedy.

Investigation of the Challenger Accident: Report of the Committee on Science and Technology. House Report 99-1016. Washington, DC: U.S. House of Representatives Committee on Science and Technology, 1986. This report's findings are similar to those of the Roger's Commission, drawing conclusions and making recommendations on the *Challenger* accident. It comments that the pressure to maintain a s chedule o f 2 4 l aunches p er y ear prompted NASA to take more shortcuts than appropriate and blames not only NASA but also Congress and the administration f or c ontributing to th is pressure, including that to evolve f rom a n R &D to a c ompetitive operational agency, which also contributes to this difficulty. The report reemphasizes the need for safety and stresses the nonroutine nature of space flight.

"James Beggs Speaks on the Future of NASA." *Science & Technology.* 24 July 1987, pp. 17-21. This interview with the former N ASA A dministrator c ontains a s ignificant d iscussion o f the agency's r ole in the development of the Shuttle. Beggs had objected to th e a ppointment o f D r. W illiam G raham as d eputy ad ministrator an d h ad, according to m any s ources, b een forced o ut of the A dministrator's j ob as a r esult. G raham h ad b een o n th e scene to issue the go decision leading to the 1986 *Challenger* disaster. Beggs discusses pointedly the accident and the investigation thereafter. He notes that the r eview w as no t t horough e nough a nd t hat t he a ccident requires further examination.

Kline, T imothy E . "Walking on W ings: C aution and Courage for Manned Space Flight." *Air University Review.* 37 (May-June 1986): 70-75. In the wake of the *Challenger* accident, this article describes the balance necessary in manned s pace f light b etween t he accep tance o f r isk an d t he co mmitment t o s afety, taking a historical perspective.

Kovach, Kenneth A., and Render, Barry. "NASA Managers and Challenger: A Profile and Possible Explanation." *IEEE*

Engineering Management Review. 16 (March 1988) : 2 -6. This article is a sociological study of NASA management and the *Challenger* disaster. Based on a series of tests conducted on NASA personnel between 1978 and 1982, the authors conclude that agency leaders are characterized by a tendency not to reverse decisions and not to heed the advice of people outside the management group.

Kubey, Robert W., and Peluso, Thea. "Emotional Response as a Cause of Interpersonal News Diffusion: The Case of the Space Shuttle Tragedy." *Journal of Broadcasting & Electronic Media.* 34 (Winter 1990) : 69 -76. This article looks at the psychological aspects of disasters reported in the media using the Shuttle accident as the principal vehicle. It asserts that there is a strong and direct correlation between the strength of emotional reaction by the public and the time spent talking and showing pictures of the accident.

Lansford, Henry. " Phoenix in Space: Rising from the Ashes to Orbit." *World & I.* 5 (December 1990) : 315 -21. Discusses the return to operational status of the Shuttle and offers some observations on the revisions to the Shuttle program in the wake of the *Challenger* accident.

Lapp, Ralph E. "$10 Billion More for Space?" *New Republic.* 26 (21 February 1970). This lengthy analysis is a thoughtful critique of the Space Shuttle by a leading opponent of the manned exploration of space. Lapp, a physicist without institutional affiliation, opposed the manned flights of Apollo and suggests here that the Shuttle is a spacecraft in search of a mission. He downplays the scientific and economic benefits of the Shuttle.

Lewis, Richard S. *The Last Voyage of **Challenger**.* New York: Columbia University Press, 1988. A follow-on to his *Voyages of **Columbia**,* this book presents in a large-size format with many illustrations the story of the tragic loss of *Challenger* in 1986.

Lewis, Ruth A., and Lewis, John S. "Getting Back on Track in Space." *Technology Review.* 89 (August-September 1986): 30 -40. This article assesses U.S. space policy in light of the *Challenger* accident, reviewing the objectives and commitment. It takes a historical view of the evolution of the manned program, assesses the payload capability of the U.S. and Soviet Union, and the NASA budget as a percentage of GNP. It finds that the U.S. has received a lot for a relatively modest outlay.

Lopez, Ramon. "Impact of Challenger Loss: Future Shuttle Flights Tied to Presidential Probe Findings." *Space Markets.* 1 (Spring 1986): 40-45. Argues that the *Challenger* disaster was caused by several unfortunate miscalculations, and that now the question of its impact on the space program must be considered. Lopez asserts that June 1987 is the earliest date that the Shuttle may be operating again, so viable alternatives should be sought in the commercial world to place satellites into orbit. Unfortunately, U.S. military vehicles cannot ease the situation since all already have assigned payloads. Questions remain, according to the author, about whether or not to build another Shuttle, and whether or not the Shuttle design needs to be recast.

Magnuson, Ed. " Putting Schedule over Safety; Despite *Challenger*, the Shuttle Program Ignores Whistle-blowers." *Time.* 1 February 1988, pp. 20-21. This short article focuses on Sylvia Robins who contends that NASA and its contractors are still wedded to Shuttle launches on an accelerated and unattainable schedule in spite of the *Challenger* accident.

Marsh, George. "Eject, Eject, Eject: The Escape Pod May Be the Most Cost-Effective Solution for the Next Generation of Spacecraft." *Space.* 4 (January-February 1988): 4-8. The *Challenger* disaster brought to light the inherent risks involved in space travel. Since then the space agencies and their major contractors have placed great emphasis on concepts and systems for escaping from space vehicles. The risk of failure is highest in the first stages of launch, when the power concentration is so large. This article examines some of the recent developments in rocket extraction systems and their potential use for the Shuttle.

Marshall, Eliot. "The Shuttle Record: Risks, Achievements." *Science.* 14 February 1986, pp. 664-66. In the aftermath of the *Challenger* accident this article assesses the history of the Shuttle from the perspective of what it has accomplished for the nation. It is a favorable record of risk versus achievement, according to the author.

McAleer, Neil. *Space Shuttle: The Renewed Promise*. Washington, DC: National Aeronautics and Space Administration, n.d. This slick booklet attempts to explain exactly what took place following *Challenger* before NASA returned to flight in 1988. Emphasis is on the Shuttle's potential and flexibility and on restoring confidence in the program. A full-color format.

McConnell, Malcolm. *Challenger: A Major Malfunction*. Garden City, NY: Doubleday and Co., 1987. This book is one of several "exposes" of NASA's Shuttle development and operations management that appeared following the *Challenger* accident. Written by a journalist and containing no scholarly apparatus, the book has on its dust jacket the phrase: "The True Story of Politics, Greed, and the Wrong Stuff." It deals with the events leading up to the decision to launch the Shuttle on 28 January 1986, emphasizing the immediate causes of the accident. McConnell highlights the pressures to launch, the objections of engineers, and the internal debates on the subject. He claims that NASA was responsible for the disaster by pressing operations officials to launch the Shuttle on January 28 so that the President could mention it in that evening's State of the Union Address. He cites as evidence conversations between a NASA public affairs person and the White House Press Office. NASA has denied this contention. He also makes much of the Utah connections of NASA administrator James Fletcher in the award of the Utah-based Morton Thiokol to make the solid rocket booster. He suggests that the Reagan administration's enthusiasm for the privatization of space and the DOD's eagerness to use the Shuttle for Strategic Defense Initiative combined to cause overwhelming pressure to launch. Most serious, he alleges that the NASA reorganization following the accident was a paper tiger carried out by the same people who had been in leadership positions beforehand.

McKean, Kevin. "They Fly in the Face of Danger." *Discover*. 7 (April 1986): 48-54. This article deals with the formal assessment of risk at NASA, emphasizing the Shuttle program and failure modes in systems.

Miller, Jon D. "The *Challenger* Accident and Public Opinion: Attitudes Toward the Space Programme in the USA." *Space Policy*. 3 (May 1987): 122-40. This article discusses the results of a survey of American attitudes toward the space program and the Shuttle. The survey was carried out in three periods: before the January 1986 *Challenger* accident, immediately afterwards, and five months later. It found that the accident strongly shifted public opinion in favor of the space program and the Shuttle. Many people expected a timely resumption of Shuttle flights, although there was a delayed recognition of the significant impact of the accident on the space program. There was a shift in public attitudes toward a more positive assessment of the benefits and costs of space exploration. Positive popular response towards funding was even more marked, something rarely found in public opinion studies.

Minsky, Marvin. "NASA Held Hostage: Human Safety Imposes Outlandish Constraints on the U.S. Space Program." *Ad Astra*. 2 (June 1990): 34-37. Assesses the length to which NASA has altered its approach and hardware used in spaceflight to ensure the safety of people aboard the Shuttle. The costs of this and validity of the human spaceflight program are assessed and still found wanting.

Moore, David H. *Setting Space Transportation Policy for the 1990s*. Washington, DC: Congressional Budget Office, 1986. This short monograph reviews the Space Shuttle policy since the inception of the program and describes the process for the 1990s. It advocates a return to flight for the system, but suggests that the Shuttle does not provide assured access to space and that expendable launch vehicles are also necessary.

Moorehead, Robert W. "America's Shuttle Returns to Space." *Progress in Space Transportation*. New York: European Space Agency, 1989. pp. 81-90. This article describes the restructuring and streamlining of the Shuttle management organization following the *Challenger* accident. It identifies the associate administrators for space flight and spells out their duties, and describes the NASA policy of assigning astronauts to management positions. It a lso c omments o n th e r ole o f th e s paceflight s afety p anel. F inally, th e a uthor d iscusses n on-managerial safety enhancement programs: the solid rocket booster changes, the Shuttle crew escape systems, and landing improvements.

NASA's Plans to Procure New Shuttle Rocket Motors. Washington, DC: U.S. House of Representatives Committee on Government O perations, 1 986. T his l engthy r eport c ontains t he he aring o f t he L egislation a nd N ational Security Subcommittee on this subject conducted on 31 July 1986 after the loss of *Challenger.*

NASA's Response t o t he C ommittee's I nvestigations of t he **Challenger** *Accident.* W ashington, D C: U .S. H ouse o f Representatives Committee on Science, Space, and Technology, 1 987. This publication contains the hearings relating to the actions of NASA following the Roger's Commission report concerning the *Challenger* accident. These were held before the committee on 26 February 1987, 100th Cong., 1st Sess.

Perrow, C harles. *Normal A ccidents: L iving w ith H igh-Risk T echnologies.* (New Y ork: B asic B ooks, 1984). T his i s a study o f t he m anagement o f t echnological i nnovations an d h ow t o make them more effective in their development. The author makes plain that a normal accident is one whose failure can be predicted with careful analysis, distinguishing between linear systems (dams), complex ones (nuclear power plants), and loose ones (most manufacturing). In tightly-controlled, high-risk systems such as spaceflight, events leading to tragedy can happen s o q uickly th at in tervention is lik ely to m ake m atters w orse. I n th ose s ystems, it is impossible to anticipate and design complex safety systems; the systems become so complex that failure probabilities are enhanced. He also describes modern management theory to create mechanisms to minimize the risks in these systems. He believes risks should be analyzed and placed into one of three categories: (1) where the risks outweigh the benefits as in nuclear power plants, abandonment is desirable; (2) where the risks are presently too high as in DNA research, efforts should be suspended until acceptable levels of risk can be attained; and (3) where efforts ar e r isky b ut can b e co ntrolled t o s ome ex tent as i n ch emical p lants an d ai r tr affic c ontrol, projects should be carefully regulated and restricted. He places spaceflight in the second category.

Petroski, H enry. *To E ngineer i s H uman: T he Role of Failure in Successful Design.* (New York: St. M artin's Press, 1985). A pre-*Challenger* book relevant for its reflections upon the relationship between engineering and risk. The author comments on the development of a special faith attached to modern technology in the public mind, and the effect of recent disasters, from Three-mile Island to Chernobyl had on that confidence. This book is not simply a chronicle of accidents, but seeks to look at the process of engineering and its creative aspects apart from its scientific ones. He notes that the design process accepts failure and seeks to test and gradually develop a system, whatever it might be, that has an acceptable level of risk to operate. He cautions that nothing is error free. He ends with a discussion of structural failures and their causes, dividing them into several categories. He notes that m any r ecent f ailures ar e n ot d ue t o en gineering b ut t o p oor co nstruction, i nferior m aterials, inadequate attention to detail, or poor management and oversight.

Results o f th e D evelopment M otor 8 Test Firing. W ashington, D C: U .S. H ouse o f R epresentatives C ommittee o n Science, Space, an d T echnology, 1 987. T his p ublication r eports t o the s ubcommittee o n S pace S cience an d Applications o n t he s uccessful p erformance o f t he r edesigned S huttle s olid r ocket b oosters at the Morton Thiokol test facility in Brigham City, Utah. Hearings were conducted on 16 September 1987, 100th Cong., 1st Sess.

Ride, S ally K . *Leadership and America's Future in Space: A Report to the Administrator.* W ashington, DC: N ational Aeronautics and Space Administration, 1987. Following the *Challenger* accident NASA reassessed its posture in the space program and commissioned several studies. This one, written by an astronaut, asks the question, where s hould N ASA h ead w ith t he s pace ef fort i n t he n ext twenty years? Essentially a study in strategic planning, a major part of this book deals with the ability to reach space efficiently, safely, and reliably. Two principal m eans ar e s uggested, an d R ide i ndicates t hat a m ix i s b est, t he S huttle an d ex pendable l aunch vehicles. These should become the centerpiece of all other endeavors for NASA, she argues.

Riffe, Daniel, and Stovall, James Glen. "Diffusion of News of Shuttle Disaster: What Role for Emotional Response." *Journalism Quarterly*. 66 (Autumn 1989): 551-56. This article is a study of the reporting (or over reporting) of the *Challenger* accident and the viability of emotional stories. It assesses the response of the public to this type of media coverage and offers some sophisticated analysis of the process.

Rogers, William P. *Report of the Presidential Commission on the Space Shuttle Challenger Accident*. Washington, DC: Government Printing Office, 1986. 5 Vols. The first volume of this publication contains the report itself, while the rest have supporting documentation and testimony. This is an exceptionally important study based on the Commission's investigation of the *Challenger* accident. It has aroused controversy in all quarters as being either too lax in its indictment of NASA's management or too harsh in its criticism.

Roland, Alex. "Priorities in Space for the USA." *Space Policy*. 3 (May 1987): 104-14. This article follows the story of the Shuttle development, placing it in the context of the history of the U.S. space program from Apollo to the Space Station. The Shuttle was, according to Roland, one of a series of space "spectaculars" and has proven to be expensive and unreliable, practical only for a very limited number of specialized missions. The space station also cannot be justified on a cost-effective basis, and the author concludes that the station and the replacement orbiter for the *Challenger* should be cancelled. In their place NASA should begin a major program to develop a new launch vehicle independent of the military. The aim should be toward a dramatic reduction in launch vehicle costs, making spaceflight practical, and a truly independent NASA, which could restore the United States to space preeminence. This article is followed by a response from John M. Logsdon and a rejoinder from Roland.

Roland, Alex. "The Shuttle's Uncertain Future." *Final Frontier*. April 1988, pp. 24-27. Written by a critic of the manned space program in general and the Shuttle in particular, this article assesses the state of the nation's space program in the two years following the *Challenger* accident. Roland contends that the process of developing the Shuttle was too politicized and cost-conscious for it to result in a reasonably safe system. More important, he maintains that despite, or perhaps because of, the Shuttle's technical sophistication, it is inherently flawed as a reliable vehicle to place cargo in orbit. He suggests that the only way out is for NASA to begin seeking alternatives to the Shuttle for launching payloads.

Sehlstedt, Albert, Jr. "Shuttle's History Provides Answers." *Baltimore Sun*. 12 October 1986, pp. 6-9. More than just a news story, this lengthy feature is a cogent analysis of the problems in the development of the Shuttle that led to the *Challenger* disaster. Sehlstedt points to the problems of political compromises on funding forcing technological compromises in the Shuttle. Ultimately they caught up with NASA.

Shayler, David. *Shuttle **Challenger***. London: Salamander Books, 1987. Another picture book, this large-format work is a discussion of the system, its performance, missions, and other assorted tidbits concerning the *Challenger*. There are descriptions of its construction, missions, and the accident. There is some discussion of the inquiry into the accident, as well as biographies of each of the astronauts flying on the spacecraft.

Space Shuttle Accident: NASA's Actions to Address the Presidential Commission Report. Washington, DC: U.S. House of Representatives Committee on Science, Space, and Technology, 1987. This contains the NASA briefing to the committee chairman on 30 October 1987.

Shuttle Recovery Program. Washington, DC: U.S. Senate Committee on Commerce, Science, and Transportation, 1988. This publication contains the hearings on this subject before the subcommittee on Science, Technology, and Space held on 16 February 1988, 100th Cong., 2d Sess. It contains an overview of the recovery program and focuses on the management of risk.

Sidey, Hugh. "Pioneers in Love with the Frontier." *Time*. 10 February 1986, pp. 46-47. This thoughtful discussion of the development of the U.S. space program emphasizes the role of the frontier and the exploration imperative in the United States. Sidey, an extremely articulate commentator, suggests that nothing worthwhile is gained without sacrifice. This was a response to the naysayers of the space program after the *Challenger* accident.

Simon, Michael C., and Hora, Richard P. "Return of the ELVs." *Space World.* January 1988, pp. 15-18. This article reports on the development and construction of a new generation of expendable launch vehicles. After the *Challenger* disaster NASA and everyone else realized the short-sightedness of disallowing access to space via expendable boosters. A crash program began to remedy this problem, the fruits of which the authors describe.

Simon, Michael C. *Keeping the Dream Alive: Putting NASA and America Back in Space.* San Diego, CA: Earth Space Operations, 1987. This is a slim volume that is chiefly interesting because of its discussion of the difficulties NASA has experienced in meeting the challenge of using and exploring space. It has two full chapters on the Shuttle and its development, as well as the effort of selling it to the public in the 1970s as the central means of access to space. There is also a discussion of *Challenger* and the difficulties created and problems illuminated by the disaster.

Space Shuttle Accident. Washington, DC: U.S. Senate Committee on Commerce, Science, and Transportation, 1986. This is a set of hearings conducted by the Senate subcommittee on Science, Technology, and Space in the 99th Cong., 2d Sess., on the accident and the Rogers Commission report.

Space Shuttle Oversight. Washington, DC: U.S. Senate Committee on Commerce, Science, and Transportation, 1987. This contains hearings held on the subject, prompted by the *Challenger* accident held on 22 January 1987 before the subcommittee on Science, Technology, and Space, 100th Cong., 1st Sess. This hearing deals specifically with the accident, NASA management, design and safety of the Shuttle, and launch operations.

Space Shuttle Recovery. Washington, DC: U.S. House of Representatives Committee on Science, Space, and Technology, 1987. This publication contains the text of hearings before the Subcommittee on Space Science and Applications at the 100 Cong., 1st Sess.

Stine, G. Harry. "The Sky Is Going to Fall." *Analog Science Fiction/Science Fact*, August 1983, pp. 74-77. In an article that in retrospect appears prophetic, the author describes several problems with the Shuttle and comments on what he considers a serious possibility that a major malfunction could destroy a mission and all aboard. He rests his argument on the complexity of the system and the inherent dangers of space flight. He assumes that there will be no way to prevent this accident—all activities of this magnitude eventually have a disaster—but what Stine wants his audience to do is to spearhead opposition to what he thinks will be an attack on the space program coming as a result of an accident. He wants to ensure that the baby is not thrown out with the bathwater. He urges everyone to blunt a media attack. He wants to save the space program because he is convinced that the wellbeing of humanity rests on exploration of the solar system.

Strategy for Safely Returning Space Shuttle to Flight Status. Washington, DC: U.S. House of Representatives Committee on Science and Technology, 1986. Text of hearings on this subject before the subcommittee on Space Science and Applications conducted on 15 May 1986, 99th Cong., 2d Sess. It emphasizes the safety issues of flight and the redesign of the solid rocket boosters before a return to flight.

Taylor, Stephen. "Aerojet in Focus." *Space.* 4 (July-August 1988): 34-37. The launch propulsion industry received a boost from the growing demand for launch services brought on by the grounding of the Shuttle after the *Challenger* accident. The emphasis is on the growth of Aerojet as a result of these developments, and it is currently aiming, with NASA, to reduce the cost of low earth orbit to $300 per pound.

Tests of the Redesigned Solid Rocket Motor Program. Washington, DC: U.S. House of Representatives Committee on Science, Space, and Technology, 1988. Text of the hearings on the booster redesign effort undertaken by

NASA through the primary contractor, Morton Thiokol, held on 27 January 1988.

Trento, Joseph J., with reporting and editing by Susan B. Trento. *Prescription for Disaster: From the Glory of Apollo to the Betrayal of the Shuttle.* New York: Crown Publishers, 1987. Not truly an investigation of the *Challenger* accident, this book is an in-depth review of the NASA management and R&D system emphasizing the agency's "fall from grace" after the Apollo program. Essentially Trento argues that the giants of the 1960s, the men who successfully managed the lunar program, were gone and were replaced with government bureaucrats who played the political game and sold the Shuttle as an inexpensive program and, in the process, sowed the seeds of disaster. Trento blames the Nixon administration for politicizing and militarizing the space program, and every NASA administrator since that time has had to play hard, but against bigger opponents, in both arenas. Declining every year since then, NASA was truly in the doldrums by the time of the *Challenger* accident. He argues that the failure was not the O-rings that ignited the spacecraft, it was the political system that produced them.

Wainright, Louden. "After 25 Years: An End to Innocence." *Life.* March 1986, pp. 15-17. With the characteristic *Life* emphasis on photographs, the longtime writer for the magazine assesses the space program after the *Challenger* accident.

"Whistle-blower." *Life.* March 1988, pp. 17-19. This is an interview with Roger Boisjoly, the former Morton Thiokol engineer who complained to the media and anyone else who would listen that his company and NASA had neglected critical safety indicators and allowed the Shuttle to be launched against many people's objections. That management decision led to the loss of the *Challenger*, millions of dollars, lots of time, an untold amount of credibility, and most important the lives of seven people. Contains numerous photographs.

Whitehead, Gregory. "The Forensic Theater: Memory Plays for the Post-Mortem Condition." *Performing Arts Journal.* 12 (Winter-Spring 1990): 99-110. This article assesses the use of traumatic shock from the death of loved ones or the immediacy of death brought to the screen in theater. As only one example, the author uses the *Challenger* disaster as a vehicle to assess psychological and collective behavior.

Wright, John C.; Kunkel, Dale; Pinon, Marites; and Houston, Aletha C. "How Children Reacted to Televised Coverage of the Space Shuttle Disaster." *Journal of Communication.* 39 (Spring 1989): 27-45. This is a complex study of the reactions of children to the reporting of the Shuttle accident. It uses sophisticated statistical methodology to measure six major variables and finds an intense reaction to the accident brought on by the anticipation of seeing a teacher teach a class from the Shuttle and watching the explosion on television.

Link to Part 2 (1992–2011), Chapter 6—*Challenger* Accident and Aftermath

See also Part 2 (1992–2011), Chapter 15—*Columbia* Accident and Aftermath

CHAPTER 8
SHUTTLE PROMOTION

Allaway, Howard. *The Space Shuttle at Work*. Washington, DC: National Aeronautics and Space Administration, 1979. This public relations publications booklet is a slick and simple discussion of the Shuttle and its potential. Allaway places emphasis on the role of the Shuttle in providing routine access to space.

Becker, Harold S. "Industry Space Shuttle Use: Considerations Besides Ticket Price." *Journal of Contemporary Business*. 7 (1978): 143-51. A promotional article that emphasizes the positive benefits of the Shuttle for deploying satellites, recovering or repairing items in space, and using the microgravity laboratory, which offered a whole range of new capabilities in space technology. The article concludes that the Shuttle, while an expensive program, has benefits far outweighing its costs.

Bova, Ben. "The Shuttle, Yes." *New York Times*. January 4, 1982, p. A23. This article discusses the place of the Shuttle in the overall exploration of space and the status of the United States among world powers. It is written by a well-known science fiction writer and futurist.

Collins, Michael. "Orbiter Is First Spacecraft Designed for Shuttle Runs." *Smithsonian*. 8 (May 1977): 38-47. This is an excellent article on the Shuttle's development and potential written by a former astronaut. Collins concludes that the Shuttle has the potential, however difficult it might be to fulfill, to open space for routine operations.

Faget, Maxime A. and Davis, H.P. "Space Shuttle Applications." *Annals of the New York Academy of Sciences*. 187 (25 January 1972): 261-82. This paper, written by one of the principal designers of the Shuttle, discusses the performance potential of the Shuttle and the high-energy transportation system deriving from it. The authors show that in addition to its cost effectiveness in earth-orbital missions, the Shuttle promises to be of major significance for future solar system exploration. Eventually, they suggest, the Shuttle will make possible the launching of large interplanetary payloads sent at high velocities to the far reaches of the solar system.

Gregory, William H. "Shuttle Opens Door to New Space Era." *Aviation Week & Space Technology*. 8 November 1976, pp. 39-43. One of many articles of the period which describe the Shuttle as a revolutionary system providing easy and cheap access to space.

Haggerty, James J. "Space Shuttle, Next Giant Step for Mankind." *Aerospace*. 14 (December 1976): 2-9. This is a general article on the Shuttle's development with a heavy emphasis on the potential of it to offer routine access to space. Its development is explicitly compared to the lunar landing of 1969.

Irvine, Mat. "Shuttlemania." *Scale Models*. 9 (July 1978): 330-35. During the latter 1970s the Shuttle program garnered something of the same type of popular interest as had the space program of the 1960s, and it sparked a good response from the model builders. This article describes the craze in that aspect of popular culture.

Lawrence, John. "The Demythification of NASA." *NASA Activities*. 21 (November/December 1990): 3-5. This article summarizes the main arguments NASA uses to counter its critics. Includes justifications of the Space Shuttle and its performance.

Lyndon B. Johnson Space Center. *Space Shuttle*. Houston, TX: John Space Center, 1975. This is a booklet describing for the public the potential of the Shuttle for the exploration of space. It emphasizes the benefits to be accrued and mentions the Shuttle contractors, analyzes the economic impact of the program, and describes the mission profile. It was reprinted a year later in a more concise and visually appealing form.

Meredith, Dennis. "It's 1985. Come with Commander Mitty and His Crew on a Routine 'Milk Run' Flight in the Space

Shuttle." *Science Digest*. 8 7 (January 1980): 52 -59. A lthough f lippantly n amed, th is a rticle d escribes something of the public hopes for the Shuttle in 1980 and its promise of providing routine access to space.

Michaud, Michael A.G. *Reaching for the High Frontier: The American Pro-Space Movement, 1972-1984*. New York: Praeger, 1986. Michaud presents a cogent history and commentary of the pro-space efforts made by voluntary organizations that arose near the end of the Apollo program. Michaud identifies the key groups, traces their origins and goals, and describes how they had a subtle but critical influence on the space policy of the nation during the formative years of Shuttle development. These groups lobbied with Congress and used publicity to support the space effort, not always with the expected results, however. Their intent was to turn ideas and a diffuse pro-space sentiment into legislation aimed at building support for the Shuttle and creating space stations and trips to Mars. This book represents the first systematic attempt to analyze the space booster efforts of the 1970s, and although a fine contribution, it should not be the final word on the subject.

Michener, James A. "Looking Toward Space." *Omni*. May 1980, pp. 57-58, 121. This fine article hits home to the heart of the American sense of pioneering and argues that the next great challenge in this arena is space. "A nation that loses its forward thrust is in danger," he comments, "the way to retain it is exploration" (p. 58). It is an eloquent and moving defense of the American space program in all its permutations.

Michener, James A. "Manifest Destiny." *Omni*. April 1981, pp. 48-50, 102-104. An outstanding reading experience, this article, by the dean of American popular novelists, encapsulates all the most cherished principles for manned space flight. It is human destiny to explore, he notes, and space is the next logical path. He also hangs much hope for this exploration on the Shuttle, commenting that "if the Space Shuttle succeeds, Americans will once again be voyaging in space after a period of six years. If it fails, the exploration of space may close down for several decades" (p. 102).

Mueller, George F. "The Benefits of Space Exploration Related to the Space Shuttle." *Interavia*. 27 (December 1972): 1335-36. This article is a very good NASA view of what was envisioned for the Shuttle at the time that it was being developed. Written by the chief of NASA's Office of Manned Space Flight, it emphasizes the boon to scientists of such projects as orbiting observatories and to commercial enterprise because of its ability to use the weightless environment to manufacture new materials. Accordingly, Mueller was seeking to describe to two important, but critical, constituencies that the Shuttle had real value.

Mueller, G eorge F . " Space S huttle—Beginning a N ew E ra i n S pace Cooperation." *Astronautics & A eronautics*. September 1972, pp. 20 -25. This is a useful article, but chiefly for its positive approach of the subject. It highlights the multinational promise of the Shuttle and the ready access to space it will provide humanity.

National Aeronautics and Space Administration. *Space Shuttle: Emphasis for the 1970s*. Washington, DC: National Aeronautics and Space Administration, 1972. This booklet for popular audiences describes the Space Shuttle as a vehicle that combines the advantages of airplanes and spacecraft, capable of repeatedly flying to space and back to earth. It could be launched vertically, powered by two solid-rocket boosters, which will be parachuted to earth for retrieval at an altitude of about 40 km. New uses of space flight are anticipated as costs decrease, turn-around times shorten, and operations become simplified. Color illustrations are included, but there are no references.

O'Leary, Michael. "Shuttling, the Ford of the Space Ways." *Air Progress*. 39 (December 1977): 38-44. A popular and popularizing article, this essay describes the general development of the Shuttle and what it means for the development of civilization by providing routine access to space.

Ragsdale, Al. "Flying the Space Shuttle." *Analog*. 97 (December 1977): 70-85. More heavily promotional than most, this article reviews the development of the Shuttle and hits hard the potential it has for opening up space to routine operations.

Robertson, Donald F. "The Space Shuttle in Perspective: Making a Good Space Shuttle Better." *NASA Activities*. 21 (November/December 1990): 6-7. Offers justifications for the Space Shuttle, praising its reliability, cost-efficiency, and technological achievements.

The Space Shuttle Adventure. Los Angeles, CA: Cheerios and Rockwell International, 1985. This is a short, 25-page booklet describing the Shuttle and its mission for young readers. It was put together as a promotional handout to capitalize on the popularity of the Shuttle.

Space Shuttle Program Overview. Washington, DC: National Aeronautics and Space Administration, n.d. This short, trifold brochure relates in words and a few illustrations the development of the Shuttle. Very informative as well as promotional, it is designed for the public.

"Space Shuttle—Vital to Man's Future." *Space World*. March 1974, pp. 4-35. This is a very positive description of what the Shuttle is intended to be and what it offers to the world. More useful as a gauge of public interest than in bringing new ideas to the study of the Shuttle.

Steinberg, Florence S. *Aboard the Space Shuttle*. Washington, DC: National Aeronautics and Space Administration, 1980. Designed for school classes to familiarize them with the Shuttle and its mission. Well-illustrated and written in a catchy style, it is a good example of the public relations material put out by the agency.

Taylor, L.B. "Shuttling into Space." *Mechanics Illustrated*. 66 (April 1970): 45-47. This is a short, general article on the method of operation of the Shuttle.

Von Braun, Wernher. "Coming . . . Ferries to Space." *Popular Science*. September 1965, pp. 68ff. This is a speculative article on the potential of space exploration with reusable craft, very similar to what became the Shuttle, for moving people and things between the earth and orbit. Written by the head of the rocket design team that put a man on the moon. It was an enormously successful piece which captured many people's imagination.

Von Braun, Wernher. "Spaceplane That Can Put You in Orbit: Space Shuttle." *Popular Science*. 197 (July 1970): 37-39. Promotes the Shuttle, emphasizing its strong potential for gaining easy access to space and discussing the possibilities of airliner-type operations.

Welch, Brian. "Musings of an Unabashed Shuttle Apologist." *NASA Activities*. 21 (November/December 1990): 20-21+. This article depicts the Shuttle as the historical culmination of aerospace triumphs stretching back to the Wright brothers.

Link to Part 2 (1992–2011), Chapter 1—General Works

Link to Part 2 (1992–2011), Chapter 3—The Decision to Build the Space Shuttle

CHAPTER 9
SCIENCE ON THE SHUTTLE, POTENTIAL AND ACTUAL

"A User's Eye-View of the Space Shuttle." *NASA Activities*. 21 (November/December 1990): 18-19. This is an interview with Dr. Charles E. Bugg concerning the Protein Crystal Growth Experiments performed on the Space Shuttle. The article discussed the advantages of using the Shuttle for these experiments, the results, and the possible medical advantages of the research.

Baker, David. "Programming the Shuttle to Future Needs." *Spaceflight*. 22 (March 1980): 137-40. This article takes a cursory look at the role of the Shuttle in the development of all manner of commodities that could benefit from a microgravity environment.

Billstein, Roger E. "International Aerospace Engineering: NASA Shuttle and European Spacelab." Unpublished paper prepared for the NASA-ASEE Summer Faculty Fellow Program, 12 August 1981. This paper was prepared to discuss the interrelationships between the NASA and the European space programs for the conduct of a Shuttle mission to launch the Spacelab. It deals largely with the policy and diplomatic history of the subject.

Bless, Robert. "Space Science: What's Wrong at NASA." *Issues in Science and Technology*. 5 (Winter 1988-1989): 67-73. Not specifically concerned with the Shuttle, that program nevertheless enters into Bless' analysis of the problems of NASA. He uses the Hubble Space Telescope as an example of how not to manage a program and concludes that the problems are "overreliance on the Space Shuttle, a predilection for big projects, and poor management."

Chesterton, T. Stephen; Chafer, Charles M.; and Chafer, Sallie Birket. *Social Sciences and Space Exploration: New Directions for University Instruction*. Washington, DC: NASA EP-192, 1988. This is an educational publication issued by NASA exploring the relationship between technology and society. It emphasizes technological change and its continuing effects on the society that produces it. As a pathbreaking technology, the Shuttle plays a large role in the discussions contained in this book. The book, designed for use by college professors and students, provides introductory material on a variety of space-related social topics to help in classroom explorations.

Committee on Space Research (COSPAR). *Environments of Planetary Bodies and the Shuttle*. New York: Pergamon Press, 1986. This book contains reports presented at the annual meeting of COSPAR held at Toulouse, France on 30 June-11 July 1986.

Covault, Craig. "Shuttle Launch of Galileo Jupiter Mission Highlights U.S. Space Science Renaissance." *Aviation Week & Space Technology*. 23 October 1989, pp. 22-24. This news story reports on the launch of the Galileo for Jupiter from the Shuttle mission in October 1989 and suggests that the Shuttle program is now emerging from the doldrums after the loss of *Challenger*.

Dooling, Dave. "Eyeing Innovative Shuttle Payloads." *Astronautics & Aeronautics*. 18 (May 1980): 18-20. This article describes some of the unique missions and science experiments projected for the Shuttle. It is not particularly unique in what it discusses but has some useful information.

Froehlich, Walter. *Spacelab: An International Short-Stay Orbiting Laboratory*. Washington, DC: National Aeronautics and Space Administration, 1983. This is an interesting short study of Spacelab, the Shuttle-based laboratory built by ESA as a cooperative venture with NASA. It is heavily illustrated and designed for a popular audience.

Get Away Special . . . The First Ten Years. Greenbelt, MD: Goddard Space Flight Center, 1989. This 40-page report describes t he o rigins a nd d evelopment o f t he uni que s cience p rogram f or th e S huttle that allows both professional an d n onprofessional ex perimenters t o g ain acce s s t o s pace. The brief history begins with the origins of the Get Away special, and tells about the milestones in its development. Most important, it presents an overview of individual customer payloads, chronologically grouped with the various Shuttle missions.

Goddard S pace F light C enter. *Final Report of the Space Shuttle Payload Planning Working Groups*. Greenbelt, MD: Goddard S pace F light C enter, Ma y 1 973. 1 0 V ols. D escribes in d etail th e in itial p lans f or th e u ses of th e Shuttle. In addition to the first v olume, w hich c ontains e xecutive s ummaries, o ther v olumes r eview th e potential payloads in the disciplines of astronautics, atmospheric and space physics, high energy astrophysics, life s ciences, s olar p hysics, co mmunications an d n avigation, ear th o bservations, ear th an d ocean physics, materials processing and space manufacturing, and space technology.

Greer, Jerry D. "Space Shuttle Large Format Camera Coverage of Areas in Africa—A Review of the Mission and the Photographs A cquired." *Geocarto I nternational*. 4 (June 1989) : 19 -33. D uring O ctober 1984 t he Shuttle *Challenger* carried a large format cartographic camera for an engineering evaluation that took an excellent set of high resolution photographs of limited areas worldwide with many in Africa. The results of this experiment are reviewed in this article, which also presents many striking photos.

Halstead, Thora W., and Dufour, Patricia A., eds. *Biological and Medical Experiments on the Space Shuttle, 1981-1985*. Washington, DC: National Aeronautics and Space Administration, 1986. This volume describes each of the biological and m edical e xperiments a nd s amples f lown o n th e S huttle d uring its m issions p rior to th e *Challenger* accident. It lists the Shuttle m issions c hronologically b y n umber a nd th en d escribes e ach experiment that took place on each mission, including such data as: flight number, experiment title, information on investigators, s ponsors, de velopers, m anagement an d integration t eam, ex periment l ocation in the Shuttle, species s tudied, o bjectives o f t he ex periment, a d escription of the experiment, conclusions, and references about the experiment for further research.

Hammel, R.L., G illiam, A.S., and W altz, D.M. "Space Processing Payloads—A R equirements O verview." *Journal of the B ritish I nterplanetary S ociety*. 30 (October 1977) : 363 -77. This article considers t he s pace p rocessing applications w ith r egard to th e u ser c ommunity o ffered b y th e Shuttle. The d evelopment of a s eries of l ow-gravity materials processing experiments, including c rystal g rowth a nd s olids, is d escribed a long w ith th e program requirements for such research. Spacelab should satisfy many of these efforts in partnership with the Shuttle. T he a uthors a lso r eview t he results of an eight-month s tudy w hich d efines a nd i nvestigates p ossible space application payloads for the Shuttle, with special attention to payload design criteria, mission planning, and analyses regarding costs and scheduling.

Hoffman, H .E.W. *The Spac e L aboratory: A E uropean-American C ooperative E ffort*. Washington, D C: N ational Aeronautics and Space Administration, 1 981. This short work, a translation of a W est German study, reviews the history of the European participation in the American Space Shuttle project. Some early work carried out in West Germany on the rocket-powered second stage of a r eusable launch vehicle system is cited, in particular wind tunnel studies of the aerodynamic and flight-mechanical behavior of various lifting body configurations in the subsonic range. Also highlighted is the development of in ternational c ooperation in th e S huttle p rogram, especially noting West German interest and expertise. Also mentioned is the U.S.'s decision to exclude Europe from participating in the design of the orbiter and the booster stage of the Shuttle.

Jerkovsky, W illiam. e d. *Shuttle P ointing of E lectro-Optical Experiments*. Bellingham, WA: S ociety of P hoto-Optical Engineers, 1981. T his book c ontains the pr oceedings of a t echnical conference held in L os Angeles, CA, on 10-13 February 1981 c oncerning th e us e of th e Shuttle for s tar tracking a nd o ther s cientific e ndeavors us ing optics.

Johnson, Rodney O., and Meredith, Leslie. eds. *Proceedings of the Space Shuttle Sortie Workshop*. Greenbelt, MD:

Goddard Space Flight Center, 1974. 2 Vols. This publication represents the work done at a conference on the Shuttle held at Goddard Space Flight Center on 31 July-4 August 1974. The first volume of the proceedings deals with policy and system characteristics, while the second contains working group reports. The proceedings describe the basic capabilities of Shuttle sortie mode and the potential uses of the Shuttle for research in individual disciplines.

Katauskas, Ted. "Shuttle Science: Is it Paying Off?" *Research and Development.* 32 (August 1990): 43-46, 48, 50, 52. The author claims in this article that "about half of the projects scheduled to fly into space in the orbiter will do so with little or no practical proof that they will work once they reach microgravity." He maintains—contrary to the evidence in several other publications listed in this section—that NASA does not keep adequate records of experiments and that incompatible experiments are packed together on the Shuttle, often ruining results.

Koelle, Dietrich E., and Butler, George V. ed. *Shuttle/Spacelab: The New Space Transportation System and its Utilization.* San Diego, CA: Univelt, 1981. This is a collection of papers relating to the development and uses of the Shuttle presented at the American Astronautical Society.

Lord, Douglas R. *Spacelab: An International Success Story.* Washington, DC: National Aeronautics and Space Administration, 1987. Spacelab was the European-developed and U.S.-operated space laboratory carried in the cargo bay of the Space Shuttle. This book details the history of this program from its conception, describing negotiations and agreements for European participation and the roles of Europe and the U.S. in system development, operational capability development, and utilization planning. More important, it reviews the joint management structure, coordination process, and the record in solving management and technical questions in an international setting. While the Shuttle comes into the book repeatedly as the vehicle carrying this system, this book is a chronological account of the Spacelab program from 1967 to 1985. It is filled with illustrations, many in color, and while it has no notes, a list of sources is included, as well as facsimile reprints of many important documents.

Lyndon B. Johnson Space Center. *Spacelab Life Sciences 1: First Space Laboratory Dedicated to Life Sciences Research.* National Aeronautics and Space Administration NP 120, August 1989. This glossy, well-illustrated publication discusses the first in a series of three Shuttle missions dedicated to studying how living and working in space affects the human body. The document reviews the effects of weightlessness on the body, describes some of the major experiments to be performed, and includes a brief description of the crew and the program management.

Mark, Hans. *The Space Station: A Personal Journey.* Durham, NC: Duke University Press, 1987. This is an insider's account of the space science policy developed in NASA during the period of germination of the space station. Although ancillary to a discussion of the station, it addresses in detail the technological debates over the method of traveling to and from the space station, including the effect of the *Challenger* tragedy. The development of the Shuttle and the relationship of it to the space station, arms control, and other topics are also considered.

Mason, J.A. *The Space Shuttle Program and its Implications for Space Biology Research.* Houston, TX: Johnson Space Center, 1972. This was originally presented as a paper at the American Institute of Biological Sciences meeting in Minneapolis in 1972. It deals with some of the potential for microgravity research using the Space Shuttle.

Morgenthaler, George W., and Burns, William J., eds. *Space Shuttle Payloads.* Tarzana, CA: Univelt, 1973. This is a publication of the American Astronautical Society. Consisting of papers by several people, it presents technical information on the various types and specifics of many payloads to be flown in the Shuttle orbiter.

Morgenthaler, George W., and Hollstein, M., eds. *Space Shuttle and Spacelab Utilization: Near Term and Long Term Benefits for Mankind.* San Diego, CA: Univelt Inc., 1978. A scientific work, this publication analyzes the uses of the Shuttle for microgravity, biomedicine, and other types of research.

Moulton, Robert R. *First To Fly*. Minneapolis, MN: Lerner Publications Co., 1983. This is an account of 18-year-old Todd N elson, w ho d esigned a n e xperiment t o s tudy t he f light of insects in o rbit. I t w as th e f irst s tudent experiment ever to fly aboard the Space Shuttle.

Murray, Bruce. *Journey into Space: The First Three Decades of Space Exploration*. New York: W.W. Norton and Co., 1989. This book is not principally concerned with the Space Shuttle, but it is discussed in some detail in the latter part of this highly personal acco unt b y a f ormer d irector o f J PL. M urray, w ho w as co ncerned w ith planetary probes, wrote that those missions were constantly challenged by the Shuttle, as NASA's dollars were poured into a development program which lagged behind schedule and over budget. He referred to the Shuttle as NASA's "sacred cow" which always had to be fed despite any other worthwhile projects that went begging. This was e specially tr ue d uring th e ear ly 1 980s w hen the S huttle was b ecoming o perational an d the Reagan administration w as i ntent o n c utting go vernment e xpenditures. I n e ssence, M urray c oncludes, t he S huttle priority ensured that the United States would have no mission to Halley's Comet.

National Academy of Sciences. *Scientific Uses of the Space Shuttle*. Washington, DC: National Academy of Sciences-National Research Council, 1974. This 214-page document surveys the missions that could be accomplished by the proposed S huttle. T he ar eas o f s cientific r esearch considered ar e: (1) at mospheric an d s pace p hysics, (2) high energy astrophysics, (3) infrared astronomy, (4) optical and ultraviolet astronomy, (5) solar physics, (6) life sciences, an d (7) p lanetary exploration. S pecific p rojects to be conducted in these b roader ar eas ar e al so defined. Also analyzed are the modes of operation of the Shuttle.

National Aeronautics and Space Administration. *Materials Processing in Space: Early Experiments*. Washington, DC, National Aeronautics and Space Administration, 1980. This study assesses some of the experimental activities relating to materials processing in orbit.

Naugle, John E. "Research with the Space Shuttle." *Physics Today*. 26 (November 1973): 30-37. This is an interesting article on the potential for research in space using the large capacity bay of the Shuttle.

Neal, V alerie. *Renewing Solar Science: The Solar Maximum Repair Mission*. Greenbelt, MD: Goddard Space Flight Center, 1984. This is a brief discussion of the successful efforts to repair the Solar Maximum satellite using the Space Shuttle.

Prouty, Clarke R., ed. *Get Away Special Experimenter's Symposium*. Washington, DC: National Aeronautics and Space Administration, 1984. This is a collection of papers delivered at a symposium on small-scale experiments for the Shuttle held 1-2 August 1984 at the Goddard Space Flight Center, Greenbelt, MD.

Science in Orbit: The Shuttle and Spacelab Experience, 1981-1986. Washington, DC: National Aeronautics and Space Administration, 1988. This contains a mission by mission analysis of the scientific experiments conducted in the Shuttle between its first orbital mission and the *Challenger* accident.

Shapland, D avid, an d R ycroft, Mi chael. *Spacelab: Research i n E arth O rbit*. C ambridge, E ngland: C ambridge University Press, 1984. This is a useful discussion of the development and flight of the laboratory built by Europeans for use aboard the Shuttle in earth orbit. It charts the twelve-year program of development through the first launch on the Shuttle in November 1983. It contains a chronicle of experiments performed in the lab and discusses some of the results. The book is highly illustrated with full-color in many places, and is designed as a readable work for the general public but without sacrificing detail and accuracy.

Space Shuttle P ayloads: H earing B efore the C ommittee on A eronautical and S pace Sciences. W ashington, DC: U .S. Senate C ommittee o n A eronautical an d S pace S ciences, 3 0-31 O ctober 1973. T his t wo-part p ublication reviews the Shuttle's possible missions in the 1980s.

Thomas, Lawrence R., and Mosier, Frances L. *Get Away Special Experiment's Symposium*. Washington, DC: National Aeronautics and Space Administration, 1988. This contains the proceedings of the 1988 symposium at Cocoa Beach, Florida, that NASA hosted for scientists interested in the unique experimental capability provided in the Shuttle.

Wilkerson, T homas D .; L auriente, Mi chael; an d S harp, G erald W . *Space Shuttle E nvironment: P roceedings of t he Engineering F oundation C onference, Space Shut tle E xperiment and E nvironment W orkshop he ld at N ew England College, Henniker, New Hampshire, U.S.A., August 6-10, 1984*. Washington, DC: The Engineering Foundation, 1985. A total of 26 presentations make up this technical publication about the Shuttle. Everything is oriented toward current programs and what they offer the world, as well as to projections for the future of the space program. There is considerable discussion of the role of the Shuttle in scientific endeavors in such areas as environmental experimentation; chemical, electronic and biological studies; particle and molecular research; and weightless and motion studies.

Winter, D avid L . "Carry-On Shuttle Payloads, o r H ow to 'Con th e System'." *Astronautics & A eronautics*. 1 5 (June 1977): 54 -56. This a rticle d iscusses th e p otential o f th e S huttle f or r easonably-priced s pace ex periments, especially with the so-called "getaway specials" being developed for use in the cargo bay.

Link to Part 2 (1992–2011), Chapter 7—The Space Shuttle and the Hubble Space Telescope

Link to Part 2 (1992–2011), Chapter 8—Science on the Space Shuttle

CHAPTER 10
COMMERCIAL USES OF THE SHUTTLE

Akin, David L. "Teleoperations, Robotics, Automation, and Artificial Intelligence: Technologies for Space Operations." *U.S. Opportunities in Space*. London, England: Space Consultants International, 1985. This article, presented at the second annual Space Business Conference in 1985, says that the development of the Space Shuttle opens the door to the potential development of space for commercial purposes. So far, these operations have focused on two separate technologies: manual, for piloted missions, and automated, for satellite missions. With recent developments, however, those dichotomies are no longer valid as a whole spectrum of possibilities is present. The two aspects of this spectrum dealt with here are the results of a two-year effort to categorize and evaluate the applications of automation, robotics, and machine intelligence systems for space programs and an overview of experimental efforts in space teleoperations, automation technology for space manipulators, and the crew scheduling system for space station use.

Ariane vs. Shuttle: The Competition Heats Up. Washington, DC: Television Digest, Inc., 1985. This short publication reviews the benefits and liabilities of launching satellites on the two principal means available, the Shuttle or the allegedly privately developed but still government subsidized Ariane expendable launch vehicle. With the competition for the satellite launch market in full-swing this book assesses how NASA and Arianespace reached their market positions in terms of service versus price.

Baker, David. "Space Shuttle: A User's Guide." *Flight International*. 20 May 1978, pp. 1552-58. A review of the Shuttle's potential, emphasizing its commercial and practical activities.

Banks, Howard. "Overloaded Shuttle." *Forbes*. 19 July 1982, pp. 33-34. This article comments on the difficulties of buying space on the Shuttle for either scientific experiments or satellite launches.

Becky, Yvan. "Commercial Use of the Space Transportation System: Toward a Permanent Manned Space Station." *Proceedings of the International Conference and Exhibition on the Commercial and Industrial Uses of Outer Space*. Montreux, Switzerland: Interavia Pub. Group, 1986. pp. 57-66. Discusses the resumption of space transportation system operations, including the strategy for a safe return to flight, planned technical modifications to the solid booster, ground rules for a first launch, and the launch target date. According to the author, Shuttle support of commercial activities in orbit includes satellite servicing, satellite retrieval, and future support of a permanent space station.

Bennett, James, and Salin, Phillip. "The Private Solution to the Space Transportation Crisis." *Space Policy*. 3 (August 1987): 181-205. The authors of this lengthy article assert that confused and short-sighted decisions dominated by political expediency have been made about the U.S. space program for the past 30 years. Overly large and ambitious systems have been chosen, resulting in the present crisis in space transportation. The history of commercial aircraft development offers an alternative example of producing a range of sizes and capabilities for a wide variety of users and shows that the space transportation industry could benefit from applying the decision-making processes used in private enterprise. The authors examine strategies for privatization of the Shuttle and conclude that policy support for the commercial launch industry must be continued. NASA must also be reoriented toward its basic research function, and more government services should be bought from the private sector.

Bimmerle, Charles F. "Manufacturing in Space: Are You Ready?" *Twenty-eighth Annual International Conference Proceedings of the American Product and Inventory Control Society*. Falls Church, VA: APICS, 1985. The author suggests that the strategy of high technology coupled with emphasis on a global economy has brought about a second industrial revolution. A critical component of that revolution has been the space program, contributing new products and technologies to make life easier on earth. The Shuttle represents an opportunity to maximize that new development. America, via NASA, is ready to collect the economic, technological, and political rewards that can be attained from manufacturing in space. This presentation outlines the history, plans, and future of the newest type of manufacturing available to the business community, microgravity.

Blahnik, James E., and Davis, James E. "Advanced Applications of the Space Shuttle." *Journal of Spacecraft and Rockets.* 11 (February 1974): 117-119. This short article describes some of the applications anticipated for the Shuttle, emphasizing the new technologies emerging from its development.

Brown, Richard L. "Avenues and Incentives for Commercial Use of Low-Gravity Environment." *Materials Processing in Space: Proceedings of the Special Conference.* Columbus, OH: American Ceramic Society, 1983. pp. 197-209. This article discusses the new technology of microgravity when applied to the production of materials. It describes the process whereby the Shuttle in orbit can be used as a laboratory for such work, and predicts that by the end of the 1980s such activities will be routine.

Campbell, Janet W. "Choosing Reliability Level for Shuttle-Carried Payloads." *Astronautics & Aeronautics.* 14 (December 1976): 38-42. This professional paper assesses the methods and makes recommendations on the nature and means of choosing payloads for the Shuttle, essentially prioritizing those that will have the greatest immediate benefits for humanity.

Covault, Craig. "Boeing Eyes Private Shuttle Operation." *Aviation Week & Space Technology.* 2 October 1978, pp. 23-25. This is a news report of studies by Boeing to assess the possibilities for the development and operation of a Shuttle by the private sector.

Divis, Dee Ann. "Commercializing the Fifth Orbiter—Can it be Done Successfully." M. A. Thesis, University of Nebraska, 1982. This unpublished study in economics reviews the feasibility of privatizing the fifth orbiter. Developed as a hypothetical situation, the scenario is played out using readily available construction and operational data. Not surprisingly, the author finds the approach viable. The most interesting aspect of this study is a discussion of the government's position on commercialization of an orbiter.

Divis, Dee Ann. "Thinking Big by Keeping it Small: The Price and Scheduling Advantages of a Fully Reusable Mini-Shuttle." *International Space Business Review.* 1 (June-July 1985): 38-43. This article comments on the efforts of Third Millennium, Inc., to develop a new design and approach to the Space Shuttle system. Its space van system is based on reusable technology and promises commercial, airline-type operations. The launch services include a seven-day turn-around, a one-month lead time, the ability to schedule additional or emergency launches, and a launch price of $1.9 million to $40 million, depending on orbit and weight. These conditions mean both small and large companies will be able to take advantage of the opportunities in space.

Fink, Donald E. "On-Orbit Satellite Servicing Explored." *Aviation Week & Space Technology.* 14 April 1975, pp. 35-39. One of the potentials of the Shuttle was always the ability either to go into space and retrieve satellites and other objects for repair or to fix them while still in orbit. This article discusses this possibility as it was being studied by NASA.

Gillam, Isaac T. "Towards Industrial Development in Space." *Space Communications Broadcast.* 5 (March 1987): 37-43. The industrial and commercial uses of space promise substantial tangible benefits for large numbers of people throughout the world, but this effort is not without risk. NASA's Office of Commercial Programs was established in 1984 in order to provide a focus for the agencywide program to encourage private investment in commercial space activities and to facilitate technology transfer. The Shuttle program has been one focus of these efforts.

Hosenball, S. Neil. "The Space Shuttle: Prologue or Postscript?" *Journal of Space Law.* 9 (Spring-Fall 1981): 69-75. This article treats the development of the Shuttle as a method for easy access to space, focusing on the problems and potential of space commercialization, the legal issues of orbiting civilians, and associated questions. As might be expected, it is heavy on policy and legal questions and short on technological discussions.

McMahan, Tracy, an d Neal, Valerie. *Repairing Solar Max: The Solar Maximum Repair mission*. Greenbelt, MD : Goddard Space Flight Center, 1984. Discusses the successful Shuttle mission to repair a satellite in orbit.

Moore, David H. *Pricing Options for the Space Shuttle*. Washington, DC: U.S. Senate Budget Committee, 1985. This government report e xplains p ricing o ptions f or th e N ASA S pace S huttle's c ommercial activities. It also analyzes Shuttle system costs, reviews alternative cost bases for pricing policy, and examines the implications of policy options for space policy objectives.

Moore, W.F., and Forsythe, C. "Buying a Shuttle Ticket." *Astronautics & Aeronautics*. 15 (January 1977): 34-40. This paper concerns a preliminary draft for reimbursement for Shuttle flights that had been developed by NASA. It comments on the reimbursement p olicy, t he t ransition f rom ex pendable t o r eusable s ystems, t he n ew u ser services, and the economics of these activities in relation to the cost of operating.

National A eronautics a nd S pace A dministration. *Operational C ost E stimates, Spac e Shut tle—Development of U ser Charge P olicy on R eusable Spacecraft*. Washington, D C: U.S. House o f R epresentatives Subcommittee on Space Science and Applications, December 1976. This report presents the rationale for NASA's estimate of out of pocket cost of $10.5 m illion (1971 dollars) per Shuttle flight if 60 missions were flown each year. Shuttle operating costs are used to develop the charge policy for various government and industrial users of the space transportation system. It also includes a comparison of various reimbursable service charges.

Projection of Non-Federal Demand for Space Transportation Services Through 2000—An AIAA Assessment for the Office of Science and Technology Policy, the White House. New York: American Institute of Aeronautics and Astronautics, 1 981. T his study as sesses the m arket for s pace t ransportation—launch, r epair, an d r ecovery of satellites p rincipally—from w hich t he S huttle m ight b enefit, f inding s ufficient d emand f or t he p rogram t o justify its continuation.

Roberge, J .L. " Health E mergency L earning P lan (H.E.L.P.)—Down-to-Earth Applications of S pace S huttle Technology." *Emergency M edical S ervices*. 8 (July/August 1979) : 11, 14, 16-17. T his ar ticle d escribes a commercial spin-off program resulting from the Shuttle effort.

Simon, E llis. " Insurance L iftoff K ey to S pace S huttle B lastoff." *Business I nsurance*. 24 (July 1978) : 128 -40. T his lengthy a rticle d iscusses t he r ole o f t he i nsurance c ommunity in p reserving the investments of organizations involved in the space program and how the Shuttle affects that program.

Williamson, Ray A. "The USA and International Competition in Space Transportation." *Space*. 3 (May 1987): 115-21. This article is one of several that appeared during the latter 1980s reviewing the problems of competition for commercial l aunches o n the S pace S huttle an d o ther l ifting v ehicles. W illiamson ex amines d evelopments i n international space transportation from 1982 to 1992 and t he f ailure of U .S. pol icies t o m eet f oreign commercial competition in space launches. Two goals have emerged from the U.S. policy debate: to achieve assured access to space and to reduce the costs of sending payloads into orbit. Both goals need to be faced within the context of a wider commitment by government and private industry to space investment.

Woodcock, Gordon R. "Rethinking Our Space Future." *Space Manufacturing 4: Proceedings of the Fifth Conference*. Princeton, NJ: Princeton University, 1981. pp. 295-99. The best way to revitalize the U.S. Space program, according t o W oodcock, i s t o f orce S huttle o perations i nto co mmercial av enues b y making them self-supporting. This would allow the exploitation of many new technologies and make feasible the placement of a space station in orbit.

Yardley, J.F. *Space Transportation System Payload Status and Reimbursement Policy*. Washington, DC: Committee on Science and Technology, 1978. In U.S. House of Representatives Report N78-12127 03-16. This 85-page study presents a status report on the space transportation system, emphasizing the management structure and project planning, use and payloads, cost assessments, and pricing policy.

Link to Part 2 (1992–2011), Chapter 9—Commercial Uses of the Space Shuttle

CHAPTER 11
THE SHUTTLE AND THE MILITARY

Davis, P.O. "Effects of Space Transportation Systems on USAF Roles and Missions." Unpublished thesis written for Air War College, Maxwell Air Force Base, AL, 1977. This paper asserts that the Shuttle, billed by NASA as an operational vehicle, raises the specter of roles and missions within both the USAF and NASA. Should NASA operate it, or should someone else? The author asserts that USAF should fly the Shuttle and ignores the non-military aspects of the program.

Draper, Alfred C.; Buck, Melvin L.; and Goesch, William H. "A Delta Shuttle Orbiter." *Astronautics & Aeronautics*. 9 (January 1971): 26-35. This is an excellent technical review of the reasons for developing a delta-wing versus a straight-wing or lifting body orbiter. The authors were engineers for the Air Force Flight Dynamics Laboratory, and their arguments contributed to the decision to change to a delta configuration, giving the military the 2000 mile crossrange capability it needed for military missions.

Finke, R.G., and Donlan, C.J. *Continuing Issues (FY79) Regarding DOD Use of Space Transportation System*. Arlington, VA: Institute for Defense Analyses, 1979. This important study outlines several key areas relating to the partnership between NASA and the DOD on the use of the Shuttle.

Finke, R.G. *Current (FY73) Issues Regarding Reusability of Spacecraft and Upper Stages for Military Missions*. Arlington, VA: Institute for Defense Analysis, 1973. This study examines the possible contributions to military space missions of the new capabilities that would be introduced by the Space Shuttle: (1) payload recovery; (2) human presence; and (3) increased payload weight and volume at lower cost. Besides the conventional expendable mode of satellite operations, new modes of retrieval for ground refurbishment and reuse, on-orbit servicing, and on-board payloads become possible. The issue of the degree of reusability of an upper stage (tug) to be developed for use with the Shuttle for high-altitude missions is also examined. Including both transportation and payload cost savings, the results of the analysis could not support, on an economic basis, military use of a reusable tug in preference to an expendable spacecraft with minimum modification and an extended lifetime.

Francis, John J. "Planning for Reusable Launch Vehicles: A New and Necessary Outlook." *Air University Review*. 19 (November-December 1967): 98-100. This is an early assessment of the need for reusable space vehicles. Oriented toward the military program and not specifically toward NASA, it nonetheless hits at the core concern of the Shuttle, economy of operations through reusable systems.

Galloway, Alec. "Does the Space Shuttle Need Military Backing?" *Interavia*. 27 (December 1972): 1327-31. This article describes the pros and cons of military support for the Shuttle. According to the author, the Department of Defense is in a difficult position because it must support a technology that it may or may not be able to use. But without that backing, the author contends, the Shuttle could not have been supported in Congress. It concludes with the observation: "As far away as first use of the vehicle may be at this time, survival of the system from attacks against funding may ultimately depend on an agreement on joint uses in the early stages of design."

General Accounting Office. *A Second Launch Site for the Shuttle?* Washington, DC: General Accounting Office, 1978. This report to Congress by the GAO discusses the pros and cons of establishing a west coast launch site for the Shuttle at Vandenberg Air Force Base, CA.

General Accounting Office. *Space Shuttle: The Future of the Vandenberg Launch Site Needs to be Determined*. Washington, DC: General Accounting Office, 1988. This report examines the cost of reactivating the Shuttle's Vandenberg launch site.

Gillette, Robert. "Space Shuttle: A Giant Step for NASA and the Military?" *Science*. 171 (12 March 1971): 991-93. Written before the formal decision to build the Shuttle, and therefore having an air of uncertainty about the direction of the program, Gillette reviews the origins and development of the Shuttle concept through 1970. He also describes some of the configuration ideas and debates the Air Force requirement for high cross-range capability. He questions NASA's commitment to ensure that DOD needs are met: "At a development cost of somewhere between $6 billion and $25 billion, the Shuttle is likely to constitute the most expensive made-to-order gift to the nation's defense by any civilian agency." Defenders argued, Gillette comments, that the military will be the Shuttle's principal user and should therefore ensure that it meets military needs.

Heiss, K.P. "Space Shuttle Economics and U.S. Defence Potentialities." *Interavia*. 31 (November 1976): 1071-73. This article looks at the cost and organizational aspects of the Shuttle and comments on the hazards and need for back-up launch sites, payload effects, funding, fleet size, and discontinuation of the use of expendable launch vehicles. Heiss notes that the Shuttle will allow more flexibility on mass and volume of payloads, as well as greater capability to retrieve and repair satellites. It has vulnerability to sabotage, blackmail or intervention and the author suggests additional launch sites as the best means of dealing with this threat.

Henry, R.C., and Sloan, Aubrey B. "Space Shuttle and Vandenberg Air Force Base." *Air University Review*. 27 (September-October 1976): 19-26. This paper discusses the problems of siting Space Shuttle launch and landing facilities, and evaluates studies of acceptable sites. It mentions the constraints of sites—population, launch azimuths, booster impact zones, buffer zones for communities, and environmental impact. The authors note that the best sites are in coastal zones, as are those that have already been developed somewhat. All of the positive features come together to point toward Vandenberg as the second Shuttle launch site.

Henry, R.C., and Sloan, Aubrey B. "Space Shuttle and Vandenberg Air Force Base." *Space World*. February 1977, pp. 29-36. This article presents a good overview of the projected use of Vandenberg as a second launch site for the Shuttle. Its use could give NASA a launch capability on either coast and the ready capability to launch polar orbits. Moreover, it would speed the recovery of an orbiter following a landing at Dryden. It is very close in content to the earlier article by the same authors.

Holder, William G. "The Many Faces of the Space Shuttle." *Air University Review*. 24 (July-August 1973): 23-35. This article discusses the Shuttle program from a general perspective and assesses its military implications.

Johnson, E.W. "Space Transportation System: A Critical Review." Unpublished thesis written for Air War College, Maxwell Air Force Base, AL, 1974. This paper assesses the potential of the Shuttle program for military use.

Mangold, S.D. "The Space Shuttle: A Historical View from the Air Force Perspective." Thesis, Air Command and Staff College, Maxwell Air Force Base, AL, 1984. This is a simple discussion of the development of the Shuttle, and the interplay of DOD and NASA in that process, written with the biases of the Air Force by a student at an intermediate service school.

Moore, James P. "Partners Today for Tomorrow: The Air Force and the Space Shuttle." *Air University Review*. 33 (May-June 1982): 20-27. This article assesses the joint aspects of the development and use of the Space Shuttle. Appearing only a few months after the first operational mission of the Shuttle, it especially reviews the military mission of the spacecraft.

Parrington, Lt Col Alan J. "Toward a Rational Space Transportation Architecture." *Airpower Journal*. 5 (Winter 1991): 47-62. This rambling article considers the military's need for a reliable system of space transportation. After reviewing the physics of satellite orbits, the article discusses the history of space transportation. It notes that compromises on the Shuttle's design have lessened its utility, and the military's continued reliance on expendable launch vehicles is shortsighted. Thus, a new space transportation system is needed that will be able to supply a military-dedicated space station in the twenty-first century.

Sloan, Aubrey B. "Vandenberg Planning for the Space Transportation System." *Astronautics & Aeronautics*. 19 (November 1981): 44-50. Reviews the Air Force's efforts to develop the facilities required to operate the Shuttle out of Vandenberg Air Force Base. It suggests, somewhat optimistically, that the first such launch will take place there in 1985.

Smith, Bruce A. "Military Space System Applications Increasing." *Aviation Week & Space Technology*. March 9, 1981, pp. 83-87. This article assesses the role of the Shuttle, and other space programs, in the defense of the United States.

Smith, Bruce A. "Vandenberg Readied for Shuttle Launch." *Aviation Week & Space Technology*. 7 December 1981, pp. 49-52. This news story deals with the activities required to make Vandenberg a suitable Shuttle launch site, and thereby broaden the options for Shuttle usage.

Steelman, Donald L. "The Air Force and the Space Transportation System." *Air University Review*. 22 (January-February 1971): 34-41. This article assesses the Shuttle program as it relates both to NASA and to the DOD.

Ulsamer, Edgar. "Space Shuttle, High-Flying Yankee Ingenuity." *Space World*. June 1977, pp. 18-23. This article discusses the Space Shuttle program from the standpoint of its potential military uses. Ulsamer points out several advantages to the program, particularly its ability to put very large antennae and power sources into space and its retrieval capability. The Shuttle's high payload capability could accelerate the development of space-qualified high-energy laser systems. The Titan III will only be used as a backup to the Shuttle launch system. The author spends considerable time discussing the attributes of the solid-propellant, expendable, high-orbit Interim Upper Stage, able to send payloads beyond geostationary orbit. It would have great capability for the 24 satellite global positioning system then being developed for USAF.

Ulsamer, Edgar. "Space Shuttle Mired in Bureaucratic Feud." *Air Force Magazine*. September 1980, pp. 72-77. Written from a decided pro-military position, this detailed article reviews the debate among NASA, the DOD, and other federal agencies over the role of the Shuttle. The real issue is whether or not the nation's space policy, open and civilian, should be militarized. The author is convinced that the military advantages of space are important enough to warrant the DOD's primacy there. He refers to those who disagree as "fuzzy thinkers" who do not understand the world, since the world, the focus of the DOD's interest, is not a safe place. Space is the new high-ground and must be exploited to keep the nation safe.

Vandenberg Space Shuttle Launch Complex. Washington, DC: U.S. Senate Committee on Commerce, Science, and Transportation, 1984. This report on the hearings before the subcommittee on Science, Technology, and Space on 10 September 1984, 98th Cong., 1st Sess., concerning construction deficiencies and quality control failures of the building effort at Vandenberg.

Wisely, Fred H. "The National Space Program and the Space Shuttle: Historical Perspectives-Future Directions." Thesis, National War College, 1981. Argues that the civilian space program under NASA has received the lion's share of the funding and publicity, while the military space program under DOD has been a backwater. This began to change as the Shuttle was developed as the "sole vehicle for future space launches." Assesses what the author considers as the three areas most important for the future: space policy, organizational structures, and hardware. In every case Wisely makes a strong argument for the primacy of the military mission in space and the need to keep those concerns paramount. He argues for a new space act that emphasizes the military aspects of the space mission. He also recommends that a single organization should be developed to manage space programs, one apart from NASA and the DOD that would operate the Shuttle for both. In terms of hardware Wisely argues against the Shuttle as the sole means to enter orbit, suggesting that expendable launch vehicles are also necessary. In the case of a Shuttle failure, he comments, the United States would have no way to launch satellites.

Link to Part 2 (1992–2011), Chapter 10—The Space Shuttle and the Military

CHAPTER 12
SHUTTLE ASTRONAUTS

"Astronauts for First Space Shuttle Flights Named." *Space World.* 7 (July 1978): 12-21. This article profiles the group of astronauts picked for the Shuttle program.

Atkinson, Joseph D., Jr., and Shafritz, Jay M. *The Real Stuff: A History of the NASA Astronaut Requirement Program.* New York: Praeger Pubs., 1985. Authors present a brief overview of the selection of the first ten groups of NASA astronauts through 1984, then concentrate on covering the watershed selections of 1959, the first group; 1965, the first scientists; and 1978, the first Shuttle selection including women and minorities. Places heavy emphasis on the criteria for selection and the procedures used, and on efforts to bring minorities and women into the Shuttle program.

Baker, David. *I Want To Fly the Shuttle.* Vero Beach, FL: Rouke Enterprises, 1988. This is a children's book on the Shuttle, describing how astronauts are chosen and trained and what it would be like to fly a mission. It is part of the "Today's World in Space" series of books that are short, highly illustrated accounts of various space exploration activities.

Bird, J.D. "Design Concepts of the Shuttle Mission Simulator." *Aeronautical Journal.* 92 (June 1978): 247-54. This article presents a solid overview of the Shuttle simulator then being developed for shuttle astronaut training.

Brandenstein, Dana nd Hartsfield, James. "Flying the Space Shuttle: A Pilot's Log." *NASA Activities.* 21 (November/December 1990): 11-15ff. Written for popular consumption, this article describes a Shuttle mission from the astronauts' standpoint.

Briefing by the Astronauts of Shuttle Mission 51-A. Washington, DC: U.S. Senate Committee on Commerce, Science, and Transportation, 1985. This publication contains the text of the presentation by the crew of Shuttle mission 51-A before the Senate subcommittee on Science, Technology, and Space, 99th Cong., 1st Sess., 28 January 1985.

Briefing by the Crew of the Space Shuttle Mission 51-D. Washington, DC: U.S. Senate Committee on Commerce, Science, and Transportation, 1985. This publication contains the text of the presentation by the crew of Shuttle mission 51-D before the Senate subcommittee on Science, Technology, and Space, 99th Cong., 1st Sess., 6 June 1985.

Catchpole, J.E. "EVA and the Space Shuttle." *Spaceflight.* 20 (May 1978): 174-75. This short article looks at the possibilities inherent for work in space and postulates some technological developments required to make EVAs practical experiences.

Challengers: The Inspiring Life Stories of the Seven Brave Astronauts of Shuttle Mission 51-L. New York: Pocket Books, 1986. This book, written by the staff of the Washington *Post,* describes the careers of the seven astronauts killed in the *Challenger* accident of January 1986. It is a relatively standard journalistic account, but it contains considerable background not found elsewhere.

Cohen, Daniel and Susan. *Heroes of the Challenger.* London: Archway Paperbacks, 1986. Published in July 1986 about the people killed in the Challenger explosion. Provides a media approach to covering the life stories and events of STS-51L in an inexpensive and compact version of *Challengers.*

Cooper, Henry S.F. *Before Life-off: The Making of a Space Shuttle Crew.* Baltimore, MD: The Johns Hopkins University Press, 1987. The first in the New Series in NASA History, this book presents a fine discussion of the s election a nd tr aining o f c rews f or in dividual S huttle missions. Written in a journalistic style without scholarly apparatus, it is an excellent first person account of the 1984 mission of STS-41G.

Covault, Craig. "Shuttle E VA S uits Incorporate A dvances." *Aviation Week & Space Technology*. 16 M arch 1981, pp. 69-73. A description of the new space suits developed for Shuttle crewmembers for EVA, as well as some discussion of the new maneuvering units under development.

Dwiggins, Don. " Flying th e S paceship ' Enterprise' S imulator." *Plane and P ilot*. 13 (March 1977): 18 -23. T his i s a personal reminiscence of the experience of flying the Shuttle simulator used for the test orbiter.

Flight of STS-7 with Astronauts Capt. Robert L. Crippen, Capt. Frederick H. Hauck, Col. John M. Fabian, Dr. Sally K. Ride, and D r. N orman E. Thagard. Washington, DC: U .S. H ouse o f R epresentatives C ommittee o n S cience and T echnology, 1 983. T his p ublication c ontains th e h earing o f th is c ommittee with the crew and other participants involved in this mission, which was notable for several reasons, among them the flight of the first American female astronaut.

Flight of STS-8 with Astronauts Capt. Richard H. Truly, Comdr. Daniel C. Brandenstein, Lt. Comdr. Dale A. Gardner, Jr., L t. C ol. G uin S. B luford, J r., and W illiam E dgar T hornton, M .D. Washington, D C: U .S. H ouse o f Representatives Committee on Science and Technology, 1 983. This publication contains the hearings of this committee conducted with participants in the mission held on 27 September 1983.

Fox, M ary V irginia. *Women Astronauts: Aboard the Shuttle*. New Y ork: J. Messner, 1984. This book, written for the youth m arket, describes the J une 1 983 flight of the Space Shuttle with emphasis on the experiences of Sally Ride, the first American woman to fly in space. It also includes brief biographies of the eight women Shuttle astronauts.

Hohler, Robert E. *"I Touch the Future . . ." The Story of Christa McAuliffe*. New York: Random House, 1986. Written by a journalist of the *Concord Monitor*, the hometown newspaper of McAuliffe, this book is a well-researched and well-written biography of the teacher killed in the *Challenger* accident. It tells how she became interested in the Shuttle and how she competed to become an astronaut.

Koch, D avid C . " Space S huttle T raining." *Air L ine P ilot*. 47 (March 1978): 13-15. A d escription o f th e in tensive training program undergone by Shuttle astronauts.

Nelson, B ill, w ith B uckingham, Jamie. *Mission: A n A merican C ongressman's V oyage to Space*. New York: Harcourt, Brace, Jovanovich, 1988. This book is a personal account of Florida Representative Bill Nelson's flight on the Shuttle *Columbia* made only 16 days before the 28 January 1986 loss of *Challenger*. Nelson, the chair of the House Space Science and Applications Subcommittee, was a payload s pecialist o n m ission S TS-61C. T his book relates his training regimen and p reparation f or th e f light a s w ell a s th e f irst-person acco unt o f th e mission. At every level, *Mission* has an "I was there" quality about it, and is entertaining and insightful in that capacity. Most interesting, the last part of the book analyzes the *Challenger* accident and examines U.S. space policy. While costly, Nelson concludes, the risks are worth what will come out of the endeavor. He concludes his final chapter with these comments: "If America ever abandoned her space ventures, then we would die as a nation, b ecoming s econd-rate in our own eyes, as well as in the eyes of the world. . . . Our prime reason for commitment can be summed up . . . space is our next frontier" (p. 296).

Peebles, C urtis. "Training for th e Space Shuttle." *Spaceflight*. 20 (November 1978): 393-95. This is a synopsis of the training activities of the Shuttle astronauts. It should be used in conjunction with Henry S.F. Cooper's *Before Lift-off*.

Siepmann, H.R., and S hayler, D .J. *From the F lightdeck 4: N ASA Spac e Shut tle*. London: I an A llen, Ltd., 1987. This book is a popularly-oriented work that follows a typical Shuttle mission from take-off to landing, observing events and the crew's routine operations from the perspective of astronauts. It contains transcripts of dialogue between the Shuttle and mission control, a lot of pictures, and a mission statement on each of the first 25 flights. There is no scholarly apparatus.

Spangenburg, Ray, and Moser, Diane. *Living and Working in Space*. New York: Facts on File, 1989. This publication contains information about the operational activities of NASA's Shuttle missions designed for an informed non-specialist audience that uses the library to track down information.

Velupillai, David "Shuttle Training: The Final Countdown." *Flight International*. 13 December 1980, pp. 2177-82. This is a short discussion of the training program for Shuttle astronauts. Should be read in conjunction with Henry S.F. Cooper's book, *Before Lift-Off*.

Link to Part 2 (1992–2011), Chapter 11—Space Shuttle Austronauts

Link to Part 2 (1992–2011), Chapter 19—Memoirs About the Space Shuttle

CHAPTER 13
THE SHUTTLE IN INTERNATIONAL PERSPECTIVE

d'Allest, Frederic. "Ariane vs. Shuttle: The State of Competition." *Air & Cosmos*. 11 March 1978, pp. 25-29. This short article surveys the development of the Shuttle in the latter 1970s as a competitor to the European Space Agency's projected heavy-payload launch vehicle, *Ariane*.

Arenstein, Seth. "Blizzard from Baikonur." *Ad Astra*. 1 (February 1989): 14-18. Describes the development of the Soviet Shuttle "Buran" and the Soviet space program, emphasizing design, construction, and space policy, and then compares it to the U.S. Shuttle program. Comments on the flight of the Buran Shuttle, launched from Baikonur on 16 November 1988.

Billstein, Roger E. "International Aerospace Engineering: NASA Shuttle and European Spacelab." Unpublished paper prepared for the NASA-ASEE Summer Faculty Fellow Program, 12 August 1981. This paper was prepared to discuss the interrelationships between the NASA and the European space programs for the conduct of a Shuttle mission to launch the Spacelab. It deals largely with the policy and diplomatic history of the subject.

Culbertson, Phillip E., and Bold, T.P. "Opening a New Era in Space—Space Transportation System Utilizing Shuttle, Spacelab, and Interim Upper Stage." *Astronautics & Aeronautics*. 15 (April 1977): 20-25. This article explores the overall payload planning efforts aimed at initial projected use of the Shuttle to establish a new capability for exploring space through operations that could not be performed before. The first payloads were suppose to fly on orbital test flights beginning in March 1979. After these tests the Shuttle was expected to build up to as many as 60 flights a year by 1984. The payloads have been chosen to make special contributions to the management on a global scale of the interrelationships of production, consumption, population growth, and pollution.

Forbrich, Carl A., Jr. "The Soviet Space Shuttle Program." *Air University Review*. 31 (May/June 1980): 55-62. A cogent review and assessment of what was known in an unclassified setting about the Soviet Union's efforts to build its own reusable orbiter with a configuration similar to the United States.

Froehlich, Walter. *Spacelab: An International Short-Stay Orbiting Laboratory*. Washington, DC: National Aeronautics and Space Administration, 1983. This is an interesting short study of Spacelab, the Shuttle-based laboratory built by ESA as a cooperative venture with NASA. It is heavily illustrated and designed for a popular audience.

Gatland, Kenneth. "A Soviet Space Shuttle?" *Spaceflight*. 20 (September/October 1978): 322-26. A solid piece reporting that the Soviet Union was apparently working on its own version of a reusable orbiter. Gatland develops the argument that the Soviet efforts to build a Shuttle were to support the orbiting of a permanent space station by the end of the 1970s. Some time after the summer of 1973 it was reported by Soviet sources in connection with the planned operation of long-life orbital stations as "man's highway to space" that it would be necessary to have transport ships making regular flights from earth to orbit and back. The favored system, it was said, would be an unoccupied first stage and a piloted stage of aircraft type that could land at an airfield upon reentry. A test vehicle was launched in a series of glide experiments from a Tupolev Tu-95 Bear. In a broadcast by Moscow radio on 11 June 1978 it was stated that the design of the Soviet Shuttle was different from the American. The craft would resemble an aircraft with delta wings. Its rear part would carry three powerful rocket engines.

Hammel, R.L., Gilliam, A.S., and Waltz, D.M. "Space Processing Payloads—A Requirements Overview." *Journal of the British Interplanetary Society*. 30 (October 1977): 363-77. This article considers the space processing applications with regard to the user community offered by the Shuttle. The development of a series of low-gravity materials processing experiments, including crystal growth and solids, is described along with the program requirements for such research. Spacelab should satisfy many of these efforts in partnership with the Shuttle. The authors also review the results of an eight-month study which defines and investigates possible space application payloads for the Shuttle, with special attention to payload design criteria, mission planning, and analyses regarding costs and scheduling.

Hoffman, H .E.W. *The Spac e L aboratory: A E uropean-American C ooperative E ffort.* Washington, D C: N ational Aeronautics and Space Administration, 1981. This short work, a translation of a West German study, reviews the history of the European participation in the American Space Shuttle project. Some early work carried out in West Germany on the rocket-powered second stage of a reusable launch vehicle system is cited, in particular wind tunnel studies of the aerodynamic and flight-mechanical behavior of various lifting body configurations in the subsonic range. Also highlighted is the development of international cooperation in the Shuttle program, especially noting West German interest and expertise. Also mentioned is the U.S.'s decision to exclude Europe from participating in the design of the orbiter and the booster stage of the Shuttle.

International Cooperation and Competition in Space. Washington, DC: U.S. House of Representatives Committee on Science and Technology, 1 984. This publication contains the proceedings of hearings on the subject given before the subcommittee on Space Science and Applications given 25 July 1984 at the 98th Cong., 2d Sess.

Jastrow, Robert, and Newell, Homer E. "The Space Program and the National Interest." *Foreign Affairs.* 50 (April 1972): 532-44. This article is not specifically related to the Shuttle, but it presents an argument in an important forum about the nature of the space program and comments on the Shuttle in relation to that program. The authors contend that the American space program sprang principally from considerations of national security and international prestige. That motivated most of its efforts in the 1960s and informed them since that time. They assert that the Shuttle is an important step forward because it represents an opportunity to move into global service as never before because of its potential to serve the needs of all humankind. While they conclude that it will be a cost effective means of reaching space, they argue that "its prime importance lies in the fact that space is an arena in which several nations are, or soon will be, engaged. The United States must maintain a presence in that arena through a manned flight program to preserve its position as a world power."

Koelle, Dietrich E ., and Butler, George V . ed . *Shuttle/Spacelab: The New Spac e Transportation Sy stem and i ts Utilization.* San Diego, CA: Univelt, 1981. This is a collection of papers relating to the development and uses of the Shuttle presented at the American Astronautical Society.

Lord, Douglas R . *Spacelab: A n I nternational Suc cess St ory.* Washington, DC: N ational A eronautics an d S pace Administration, 1987. Spacelab was the European-developed and U.S.-operated space laboratory carried in the cargo bay of the Space Shuttle. This book details the history of this program from its conception, describing negotiations an d ag reements f or E uropean p articipation an d t he r oles o f E urope and the U.S. in system development, operational capability development, and utilization planning. More important, it reviews the joint management structure, coordination process, and the record in solving management and technical questions in an international setting. While the Shuttle comes into the book repeatedly as the vehicle carrying this system, this book is a chronological account of the Spacelab program from 1967 to 1985. It is filled with illustrations, many in color, and while it h as no notes, a list of sources is included, as well as facsimile reprints of many important documents.

Lyndon B . J ohnson S pace Center. *Spacelab L ife Sc iences 1: F irst Spac e L aboratory D edicated t o L ife Sc iences Research.* National Aeronautics and Space Administration NP 120, August 1989. This glossy, well-illustrated publication d iscusses th e f irst in a s eries o f th ree S huttle m issions d edicated t o s tudying ho w living and working in space affects the human body. The document reviews the effects of weightlessness on the body, describes some of the major experiments to be performed, and includes a brief description of the crew and the program management.

Morgenthaler, George W., and Hollstein, M., eds. *Space Shuttle and Spacelab Utilization: Near Term and Long Term Benefits for Mankind.* San Diego, CA: Univelt Inc., 1978. A scientific work, this publication analyzes the uses of the Shuttle for microgravity, biomedicine, and other types of research.

Mueller, George F. "Space Shuttle—Beginning a New Era in Space Cooperation." *Astronautics & Aeronautics.* September 1972, pp. 20-25. This is a useful article, but chiefly for its positive approach of the subject. It highlights the multinational promise of the Shuttle and the ready access to space it will provide humanity.

Piotrowski, John L. "The Challenge in Space." *U.S. Naval Institute Proceedings.* 116 (February 1990): 32-39. The author argues that "the Soviets have an impressive array of satellites, a Shuttle . . . and antisatellite capabilities that clearly display their designs for space. The United States needs to develop a space strategy and attain the capability to execute that strategy successfully."

Science in Orbit: The Shuttle and Spacelab Experience, 1981-1986. Washington, DC: National Aeronautics and Space Administration, 1988. This contains a mission by mission analysis of the scientific experiments conducted in the Shuttle between its first orbital mission and the *Challenger* accident.

Shapland, David, and Rycroft, Michael. *Spacelab: Research in Earth Orbit.* Cambridge, England: Cambridge University Press, 1984. This is a useful discussion of the development and flight of the laboratory built by Europeans for use aboard the Shuttle in earth orbit. It charts the twelve-year program of development through the first launch on the Shuttle in November 1983. It contains a chronicle of experiments performed in the lab and discusses some of the results. The book is highly illustrated with full-color in many places, and is designed as a readable work for the general public but without sacrificing detail and accuracy.

Vasil'yev, V., and Leonidov, L. *Space Transport Vehicle: Purpose, Requirements and Problems.* Washington, DC: NASA Technical Translation TT F-14, 26 July 1972. This translation of a Soviet technical paper explores the possibility of a manned, winged, reusable space transporter.

Williamson, Ray A. "The USA and International Competition in Space Transportation." *Space.* 3 (May 1987): 115-21. This article is one of several that appeared during the latter 1980s reviewing the problems of competition for commercial launches on the Space Shuttle and other lifting vehicles. Williamson examines developments in international space transportation from 1982 to 1992 and the failure of U.S. policies to meet foreign commercial competition in space launches. Two goals have emerged from the U.S. policy debate: to achieve assured access to space and to reduce the costs of sending payloads into orbit. Both goals need to be faced within the context of a wider commitment by government and private industry to space investment.

Link to Part 2 (1992–2011), Chapter 12—The Space Shuttle and International Relations

See also Part 2 (1992–2011), Chapter 16—The Space Shuttle and the *Mir* Space Station

See also Part 2 (1992–2011), Chapter 17—The Space Shuttle and the International Space Station

CHAPTER 14
MANAGEMENT AND FUNDING OF THE SHUTTLE PROGRAM

Ames, Milton B. Jr., *et al*. "Report of the Ad Hoc Subpanel on Reusable Launch Vehicle Technology." September 14, 1966. Copy in National Aeronautics and Space Administration Reference Collection, NASA History Office, Washington, DC. This report, prepared at the behest of the Aeronautics and Astronautics Coordinating Board for submission to the Supporting Space Research and Technology Review Panel, discusses the previous studies made and assesses future space mission requirements and related launch rates on a prospective basis for both DOD and NASA through the 1980 time period.

Assured Access to Space During the 1990s. Washington, DC: U.S. House of Representatives Committee on Science and Technology, 1986. This lengthy publication contains text of the joint hearings on the subject before the subcommittee on Space Science and Applications and the subcommittee on Research and Development, held 23-25 July 1985, 99th Cong., 1st Sess.

Baker, David. "The NASA Budget: Fiscal Years 1979-1980." *Spaceflight*. 21 (August-September 1979): 338-48. This article assesses the effect of the federal budget on the nation and how that relates to the space program. Baker pays special attention to the Shuttle program as the primary activity in NASA. He contends that many important objectives in space science and technology planned for the 1980s will be frustrated due to tight NASA budgets. He also suggests that more money is required to see the Shuttle program over its development hurdles and this will have an added impact on the rest of the NASA effort.

Bell, M.W. Jack. "Advanced Space Transportation Requirements and Options." *Journal of the British Interplanetary Society*. 37 (December 1984): 531-36. The author comments that the United State's space transportation system is maturing and should remain operational for the rest of the century. The use of expendable and semi-expendable elements, the massive sustaining manpower, and the required ground equipment and facilities have all contributed to a higher cost per flight than was expected. Bell advocates the construction of a new generation Shuttle that is fully reusable, lightweight, highly reliable, and equipped with long-life hardware. He believes these features can be incorporated into a single-stage-to-orbit system. This article characterizes several possible configurations for this Shuttle and illustrates some desired features. He comments, however, that Shuttle requirements are constantly evolving. The follow-on system should be defined clearly in terms of missions and applications within the limits of transportation costs. He argues, however, that a new Shuttle could not be built until a source of leadership and inspiration to initiate and lead this effort appears. While the technology is present, the will and the concomitant dollars are not.

Byrnside, N.C. "Space Shuttle Integrated Logistics: Fact or Fiction?" Unpublished thesis written for Air Command and Staff College, Maxwell Air Force Base, AL, 1979. This paper takes issue with the NASA assessment of the Shuttle's integrated logistics system, comparing it to the USAF program for supply and maintenance of weapons systems.

Carrillo, Manuel J. *A Development of Logistics Management Models for the Space Transportation System*. Santa Monica, CA: Rand Corp., 1983. This study reviews procedures and sets priorities and policies for the support of Shuttle operations.

Covault, Craig. "Space Shuttle Funding Seen at Stake." *Aviation Week & Space Technology*. 22 September 1975, pp. 47-50. This article reports on the NASA budget problems associated with the Shuttle as it ran into delays and cost overruns in the mid-1970s.

Dawson, Harry S. *Review of Space Shuttle Requirements, Operations, and Future Plans*. Washington, DC: U.S. House of Representatives Committee on Science and Technology, 1984. This report deals with the past and prospects for the Shuttle during its early operational life. It is optimistic but still not sanguine that NASA would be able to make it cost effective. This report was prepared by the House Subcommittee on Space Science and Applications.

Disher, John H. "Space Transportation: Reflections and Projections." in Durant, Frederick C., III, ed. *Between Sputnik and the Shuttle: New Perspectives on American Astronautics*. San Diego, CA: American Astronautical Society, 1981. pp. 199-224. This article is part of a larger publication focusing on various aspects of the space program. A presentation by the director of Advanced Programs for NASA's Office of Space Transportation Systems at the AAS, it contains no notes or other scholarly apparatus. It does survey the methods of spaceflight for piloted missions since Mercury and describes some of the features of the Shuttle.

General Accounting Office. *Issues Concerning the Future Operation of the Space Transportation System*. Washington, DC: General Accounting Office, 1982. This study attempts to clarify many of the problems that had arisen in the Shuttle program, especially its slower than expected mission schedule, and assesses their impact for the Congress.

General Accounting Office. *NASA Must Reconsider Operations Pricing Policy to Compensate for Cost Growth on the Space Transportation System*. Washington, DC: General Accounting Office, 1982. This report to Congress deals with the operational costs of the Shuttle and calls for a review and repricing of services provided to users of the system.

General Accounting Office. *Space Shuttle: External Tank Procurement Does Not Comply with Competition in Contracting Act*. Washington, DC: General Accounting Office, 1988. This report examines whether NASA complied with the Competition in Contracting Act for the manufacturing and fabricating of external tanks for the assembly of solid rocket boosters. It finds several deficiencies in the NASA approach.

General Accounting Office. *Space Shuttle Facility Program: More Definitive Cost Information Needed*. Washington, DC: General Accounting Office, 9 May 1977. This study looks at the costs of ground support systems and criticizes NASA for not managing the support program as carefully as thought advisable. It argues that the commitment and current estimates of NASA are not sufficiently supported by documentation, and that the facilities of the agency needed for the program have not been accurately determined. It asked that Congress require NASA to provide more definitive information from which the progress of its major facility acquisition programs could be measured and assessed.

General Accounting Office. *Space Transportation System: Past, Present, Future*. Washington, DC: General Accounting Office, 27 May 1977. This lengthy 86-page study assesses the status of NASA's Space Shuttle development program, focusing on its proposed policy for charges to users and offering several options to Congress on the question of production of orbiters in fiscal year 1978. It details the advantages and disadvantages of starting the production of a third orbiter in FY 1978 and of delaying funding of the remaining two proposed orbiters.

General Accounting Office. *Status and Issues Relating to the Space Transportation System*. Washington, DC: General Accounting Office, 21 April 1976. This study assesses NASA's Shuttle development plan and concludes that it could result in increased costs, schedule delays, and performance degradation that were not originally envisioned. The development plan, revised as the program fell behind schedule and took funding cuts, embodied such factors as reduced testing, compressed schedules, and concurrent development and production. The study also asks, but does not truly answer, whether the Shuttle system fulfills the space transportation needs of the United States.

Grey, Jerry. "Case for a Fifth Shuttle and More Expendable Launch Vehicles." *Astronautics and Aeronautics*. 19 (March 1981): 22-26. This article assesses the argument in favor of a fifth orbiter to increase the Shuttle capability as well as the development of an expendable launch vehicle for use in handling many unmanned missions. The article is prophetic in that, while it was opposed to NASA policy at the time, at least in terms of expendable vehicles, it says that the United States should not allow the Shuttle to dictate its entrance into space. This was a position especially popular after the *Challenger* accident.

Hale, Carl W. "Pricing of NASA Space Shuttle Transportation System Cargo." *Engineering Economist.* 24 (Spring 1979): 167-97. This complex article discusses the system of charges and payment for the launch of satellites and the execution of experiments by the Shuttle once it became operational. In this program, commercial firms, research institutions, and governments paid NASA for the deployment of their assets in space.

Hechler, Ken. *Toward the Endless Frontier: History of the Committee on Science and Technology, 1959-1979* (Washington, DC: U.S. House of Representatives, 1980). This book contains the best account to date of Congressional wrangling over the Shuttle and demonstrates the bipartisan nature of both Shuttle support and opposition.

Hosenball, S. Neil. "The Space Shuttle: Prologue or Postscript?" *Journal of Space Law.* 9 (Spring-Fall 1981): 69-75. This article treats the development of the Shuttle as a method for easy access to space, focusing on the problems and potential of space commercialization, the legal issues of orbiting civilians, and associated questions. As might be expected, it is heavy on policy and legal questions and short on technological discussions.

Mandell, Humboldt C., Jr. "Assessment of Space Shuttle Cost Estimating Methods." Ph.D. Diss., University of Colorado at Denver, 1983. This scholarly work reviews the methodology for arriving at costing of Shuttle components. It is a complex study, without a good story line to it, but it is integral to understanding the development of the Shuttle, especially in view of the cost-effective strategy of funding that NASA was forced to pursue in the program.

Mandell, Humboldt C., Jr. "Management and Budget Lessons: The Space Shuttle Program." *NASA SP-6101 (02),* Autumn 1989. pp. 44-48. A condensation of Mandell's Ph.D. dissertation, this article assesses the Shuttle management program and offers several recommendations. Among the most important is a concern that the program planning process is essential, requiring long and realistic planning and budget forecasting. Mandell also found that NASA needs to pay more attention to the budgeting process to ensure that sufficient funding is available during peak periods of a program, and must not accede to the demands to underestimate costs to sell the program to Congress. He concluded that NASA has a top-heavy management approach with too many large program offices at various levels of organization.

Menter, Martin. "STS—Legal Connotations." *Akron Law Review.* 13 (Spring 1981): 629-647. A really fine rundown on the legal aspects of the Shuttle: accident liability, international law, and space territorial integrity.

Mueller, George E. Address on the Space Shuttle before the British Interplanetary Society, University College, London, England. August 10, 1968. Copy in National Aeronautics and Space Administration Reference Collection, NASA History Office, Washington, DC. This presentation, made by NASA's Associate Administrator for Manned Space Flight, may well have been the first public presentation of the Shuttle concept to a scholarly community. It set up the rationale, technological choices, and planning activities taking place at NASA for the development of the Space Transportation System.

National Research Council. *Assessment of Constraints on Space Shuttle Launch Rates.* Washington, DC: National Academy Press, 1983. This is a detailed study of the ability of NASA to launch the Shuttle in the timely fashion promised to meet mission requirements. It surveys everything from the orbiter to the launch facilities to arrive at conclusions that point toward fewer launches per year than earlier anticipated. One of the important concerns of this report was the shutdown of the Shuttle production line and the hazards it had for the system's cost-effectiveness.

Office of Technology Assessment. *Reducing Launch Operations Costs: New Technologies and Practices*. Washington, DC: U.S. Congress, Office of Technology Assessment, 1988. This study reviews the practices and potential for cutting the cost of shuttle missions.

Office of Technology Assessment. *Round Trip to Orbit: Human Spaceflight Alternatives Special Report*. Washington, DC: U.S. Congress, Office of Technology Assessment, 1989. A detailed assessment of the state of the human spaceflight program and the Shuttle effort. It lays out well many of the issues affecting space policy at the end of the 1980s, e.g. whether to continue with the Shuttle in essentially its present form with minor improvements, to make major modifications, to develop a new launch system, or to develop and fly more unmanned launch vehicles.

Pace, Scott. "US Space Transportation Policy: History and Issues for a New Administration." *Space Policy*. 4 (November 1988): 307-18. The newly elected Bush administration faced complex questions on the future role of the United States in space, and tough decisions on how to pay for it. Pace comments that decisions made now on space transportation will have an important affect on U.S. space leadership in the next decade. He describes the history and current state of space transportation planning, and considers key issues that will confront the Bush Presidency. In this process the Shuttle emerges as both the cause and the effect of policy formulation.

Perrow, Charles. *Complex Organizations*. (New York: Random House, 1979). A general study, this work investigates the management of sophisticated organizations such as NASA, and offers some general insights into the way in which the Shuttle program was handled.

Pross, Mark A. "The National Aerospace Plane." *GAO Journal*. Winter 1988-1989. pp. 54-59. This article describes the NASP program and its goal "to develop and then demonstrate in a manned experimental flight vehicle—the X-30—the technologies necessary for future operational hypersonic airplanes and/or single-stage-to-orbit space launch vehicles that could deliver payloads into orbit more quickly, reliably, and inexpensively than today's Space Shuttle."

Review of the Aerospace Safety Advisory Panel Report for NASA Fiscal Year 1990 Authorization. Washington, DC: U.S. House of Representatives Committee on Science, Space, and Technology, 1990. This work publishes the hearings of the Subcommittee on Space Science and Applications held on 28 September 1989 concerning the NASA budget.

Rubenstein, S.Z. "Managing Projects— An Industry View." in *Issues in NASA Program and Project Management*. Autumn 1989. pp. 13-23. This article reviews the fundamentals of program management ala management 101, but it offers some specific lessons for the Shuttle program. It emphasizes learning from past successes and from past mistakes. It advocates the use of technology to ensure good communication between workers at all levels and tries to find a way to insulate people responsible for programs from the problems of micro-management made possible by the rapid communications medium.

Scheffer, Jim. "Shuttle Setbacks Challenge Engineers' Ingenuity." *Space World*. May 1980, pp. 14-19. This article explains better than most the reasons behind the delays in the Shuttle's development.

Shaver, R.D.; Dreyfuss, D.J.; Gosch, W.D.; and Levenson, G.S. "The Space Shuttle as an Element in the National Space Program." The Rand Corporation, October 1970. Document in the NASA Historical Reference Collection, NASA History Office, Washington, DC. This report concentrates on the economic justification and potential funding problems of the Space Shuttle. The authors expected that by 1990 the Shuttle would cost $75 billion to $140 billion while saving only $2.8 billion in space transportation costs. They predicted that satellite redesign to make optimum use of the Shuttle might result in further savings of $150 million to $200 million per year.

The report emphasizes that, due to the complexity of U.S. space transportation needs, criteria other than cost should be used to evaluate the space transportation system as then conceived.

Shuttle Derivative Vehicles Study: Operations, Systems, and Facilities. Seattle, WA: Boeing Aerospace Corp., 1977. This study deals with an assessment of unmanned cargo launch vehicles using solid rocket boosters to determine (1) vehicle concept definition, operations, and facility requirements, (2) advanced technology areas which have potential payoff in reducing operating cost, and (3) an implementation plan for a low life cycle cost system.

Space Shuttle 1975: Status Report for the Committee on Science and Technology. Washington, DC: U.S. House of Representatives Committee on Science and Technology, February 1975. This publication reports on the status of the Shuttle program before the 94th Cong., 1st Sess.

Space Shuttle 1976: Status Report for the Committee on Science and Technology. Washington, DC: U.S. House of Representatives Committee on Science and Technology, October 1975. This publication also reports on the status of the Shuttle program before the 94th Cong., 1st Sess.

Space Shuttle 1977: Status Report for the Committee on Science and Technology. Washington, DC: U.S. House of Representatives Committee on Science and Technology, 1976. This publication reports on the status of the Shuttle program before the 94th Cong., 2d Sess.

Space Shuttle 1980: Status Report for the Committee on Science and Technology. Washington, DC: U.S. House of Representatives Committee on Science and Technology, 1980. This publication reports on the status of the Shuttle program before the 96th Cong., 2d Sess.

Space Shuttle and Galileo Mission. Washington, DC: U.S. Senate Committee on Appropriations, 1980. This contains the hearings before a subcommittee of the Senate Appropriations Committee, 96th Cong., 1st Sess.

Space Shuttle Appropriations for Fiscal Year 1979. Washington, DC: U.S. House of Representatives Committee on Appropriations, 1978. This contains the hearings before the Committee of Appropriations, 95th Cong., 2d Sess. on the Shuttle.

Space Shuttle Main Engine Development Program. Washington, DC: U.S. Senate Commerce, Science, and Transportation Committee, 1978. This is the report of a hearing on the Shuttle's engine development difficulties before the 95th Cong., 2d Sess.

Space Shuttle Program: Cost, Performance, and Schedule Review. Washington, DC: U.S. House of Representatives Committee on Science and Technology, 1984. This is a transcript of the hearings held before the subcommittee on Space Science and Applications of the 98th Cong., 2d Sess.

Space Shuttle Reprogramming. Washington, DC: U.S. Senate Appropriations Committee, 1978. Text of the hearings on the subject between the subcommittee on HUD-Independent Agencies for the 95th Cong., 1st Sess.

Space Shuttle-Skylab 1973: Status Report for the Committee on Science and Astronautics. Washington, DC: House Committee on Science and Astronautics, January 1973. This publication relates hearings held on this subject by the 93d Cong., 1st Sess., and reached several conclusions about the viability of the program. It recommended that the project proceed.

Space Shuttle-Skylab: Manned Space Flight in the 1970's Status Report for the Subcommittee on NASA Oversight. Washington, DC: House Committee on Science and Astronautics, January 1972. This publication contains text of hearings held on this subject by the 92d Cong., 2d Sess., and reached several conclusions about the viability of the program, recommending that the project proceed.

Space Shuttle, Space Tug, Apollo-Soyuz Test Project-1974. Washington, DC: House Committee on Science and Astronautics, February 1974. Report of hearings held on this subject by the 93d Cong., 2d. Sess., which reached several conclusions about the viability of the program. Recommends that the program continue, noting the adequate funding for FY75 was critical to its success.

Space Tug-1973—Impact and Management of Space Tug Development Program. Washington, DC: U.S. House of Representatives Committee on Science and Astronautics, September 1973. Based on hearings during the 93d Congress, 1st Session, this document deals with the proposed space tug as a part of the Shuttle program. Congress asked for (1) a determination and finding of the role of the space tug in the Shuttle program; (2) the fiscal impact of the space tug on the overall manned spaceflight program; (3) the operations impact of the space tug on the latest proposed NASA mission model; (4) the operations impact on the project cost-per-flight of the Shuttle; (5) the ascertainment of the DOD role in the development and use of the tug; and (6) NASA's and industry's progress in conceptual design of the space tug vehicle. In concert with the above goals, the subcommittee held hearings with NASA, General Dynamics, McDonnell Douglas, Lockheed, Martin Marietta, and Grumman. As contained in the body of the hearing, NASA presented an overview of the role of the space tug program with neither conclusions nor recommended approaches.

Stevenson, A.E. "The Space Shuttle and Congress—A National Commitment." *AIAA Student Journal.* 17 (Winter 1979-1980): 4-8. As the principal element of the reusable space transportation system, the Shuttle will remain the main objective of the U.S. space program for the next twenty years. This paper analyzes various factors affecting the development of the Shuttle program with particular reference to budgeting requirements and respective congressional actions. Among the development problems that required additional funding were failures in the main engine components, particularly turbopumps, during ground tests and delays in the installation of the reusable surface insulation.

Toner, Mike. "It's Pay Off or Perish for the Shuttle." *Science Digest.* May 1985, pp. 64-67, 87-88. This article is a critical assessment of the Shuttle program written not long before the loss of *Challenger.* It notes that the Shuttle has proven to be neither cheap nor reliable, both primary selling points, and adds that the American public has been hornswoggled. Toner adds that the "concept of cost recovery is one of the legacies of the budget-conscious age in which the Shuttle program was conceived. After spending $24 billion to go to the moon, Congress wanted the Shuttle to pay its own way." That is a tall and ultimately an unfair order. NASA, the author concludes, made a blunder by accepting the cost effectiveness line instead of making the Shuttle a national resource.

Link to Part 2 (1992–2011), Chapter 13—Management of the Space Shuttle Program

CHAPTER 15
JUVENILE LITERATURE

Baker, David. *I Want To Fly the Shuttle*. Vero Beach, FL: Rouke Enterprises, 1988. This is a children's book on the Shuttle, describing how astronauts are chosen and trained and what it would be like to fly a mission. It is part of the "Today's World in Space" series of books that are short, highly illustrated accounts of various space exploration activities.

Barrett, Norman S. *Space Shuttle*. New York: Franklin Watts, 1985. This short work introduces the orbiter, what it is like inside, and what the crew does. It was written for a juvenile audience.

Barrett, Norman S. *The Picture World of Space Shuttles*. New York: Franklin Watts, 1990. A children's book, this heavily-illustrated work describes how a Shuttle works and describes various missions.

Biel, Timothy L. *The **Challenger***. San Diego, CA: Lucent Books, 1990. This book examines the *Challenger* from a scientific and historical perspective and discusses its effect on the Shuttle program. The book was written for young readers but does have a bibliography and index, as well as a good feel for the tragedy of the mission but a generally positive view toward continued space missions.

Branley, Franklyn M. *Columbia and Beyond: The Story of the Space Shuttle*. New York: Collins, 1979. This book, while written for a young audience, contains a useful general discussion of the *Columbia* orbiter, the space laboratory that fit into the cargo bay, Spacelab, and several of the experiments being developed for the Shuttle.

Cave, Ronald G, and Cave, Joyce. *Space Shuttle*. New York: Gloucester Press, 1982. This book, for a young audience, describes the general development and use of the Shuttle.

Chester, Michael. *Let's Go on a Space Shuttle*. New York: Putnam, 1975. This book, written for children, describes the Space Shuttle and suggests that in the future it will be routinely used to transport people and equipment to and from space, bring data to Earth, and rescue and repair other spacecraft.

Civil Air Patrol. *Space Shuttle: A Space Transportation System Activities Book*. Washington, DC: Center for Aerospace Education Development, n.d. This is a children's book, mostly for coloring, but with puzzles and descriptions of the Shuttle and what it will do for space exploration.

Coombs, Charles I. *Passage to Space: The Shuttle Transportation System*. New York: William Morrow, 1979. This is yet another of the run-of-the-mill books written for youth describing the construction, launching, reentry, and versatility of the Shuttle as the first reusable spacecraft.

Cross, Wilbur. and Cross, Susanna. *Space Shuttle*. Chicago: Children's Press, 1985. This is another children's book about the Shuttle.

Culper, Felix, and Peterkin, Mike. *The Infinite Beyond*. New York: Warner Juvenile Books, 1989. This is a young children's pop-up book about the Shuttle.

Dwiggins, Don. *Flying the Space Shuttle*. New York: Dodd, Mead, 1985. This is written for a young audience, describing the history of the Shuttle, its uses, how it works, and the details of a typical flight.

Dwiggins, Don. *Into the Unknown: The Story of Space Shuttles and Space Stations*. San Carlos, CA: Golden Gate Junior Books, 1971. This 80-page book, designed for young readers, describes the planning and building of the experimental space station, Skylab, and the potential of Space Shuttles for interplanetary travel.

Fichter, George S. *The Space Shuttle*. New York: Franklin Watts, 1981, 1990 rev. ed. This is a short book written for a youthful audience that describes the conception of the Shuttle, its construction, its functions, and its potential for space travel.

Fox, Mary Virginia. *Women Astronauts: Aboard the Shuttle*. New York: J. Messner, 1984. This book, written for the youth market, describes the June 1983 flight of the Space Shuttle with emphasis on the experiences of Sally Ride, the first American woman to fly in space. It also includes brief biographies of the eight women Shuttle astronauts.

Friskey, Margaret. *Space Shuttles*. Chicago: Children's Press, 1982. This is one of the many general books on the Shuttle written for children, well illustrated but otherwise pedestrian. It describes the operation and uses of a reusable system.

Jay, Michael. *Space Shuttle*. New York: Franklin Watts, 1984. Written for children, this book explains how the Shuttle works and describes a Shuttle trip from lift-off to touchdown.

Lewis Research Center. *Launching a Dream: A Teacher's Guide to a Simulated Space Shuttle Mission*. Cleveland, OH: Lewis Research Center, 1989. This publication was designed as a NASA educational publication for the use of teachers. It describes the Shuttle and a typical mission, as well as providing several resource activities for the students.

McCarter, James. *The Space Shuttle Disaster*. New York: Bookwright Press, 1988. This short book, written for juveniles, discussed the disaster that destroyed the *Challenger*, the history and possible future of Space Shuttles and the efforts of NASA to correct deficiencies in the system.

McDonald, Suse. *Space Spinners*. New York: Dial Books for Young Readers, 1991. This is a children's novel about two spiders who sneak aboard the Space Shuttle so that they can be the first spiders in orbit and to spin a web in space. Has some interesting discussion about the problems of spinning webs in a weightless environment.

McDonnell, Janet. *Space Travel: Blast-off Day*. Elgin, IL: Child's World, 1990. A children's book, this work describes the Shuttle and its mission profile.

Moche, Dinah L. *If You Were An Astronaut*. New York: Golden Book, 1985. This book, written for children, contains photographs and a simple text describing the activities of astronauts aboard the Space Shuttle.

Moore, Patrick. *The Space Shuttle Action Book*. New York: Random House, 1983. This book, describing the Shuttle for children, features pop-up and pull-tab gimmicks to follow the Shuttle on a mission from launch to landing.

Petty, Kate. *Space Shuttle*. New York: Franklin Watts, 1984. This is yet another children's book about the Shuttle, illustrated with color and describing in general terms the Shuttle's design and performance.

Redmond, Marilyn. *Henry Hamilton in Outer Space*. Gretna, LA: Pelican Pub., 1991. This is a delightful children's book about Confederate ghost Henry Hamilton. A sequel to *Henry Hamilton, Graduate Ghost*, this fictional work has Henry on vacation in Florida where he stows away on the Shuttle and haunts the astronauts in orbit.

Ride, Sally, and Okie, Susan. *To Space and Back*. New York: Lothrop, Lee, and Shepard, 1986. This book, written for younger readers, describes the experiences of space flight by the first American woman in orbit.

Ross, Frank X., Jr. *The Space Shuttle: Its Story and How to Make a Flying Paper Model*. New York: Lothrop, Lee, and Shepard Co., 1979. This book is generally useful, even if it is designed for children and leaves out much technical information. It contains an introduction to the Shuttle, its history, the construction of its major systems, a profile of a typical mission, and what the orbiter might mean to the future of space exploration. Then it provides detailed instructions on the construction of a model of the spacecraft from paper which can fly on a string or in free flight.

Steinberg, Florence S. *Aboard the Space Shuttle*. Washington, DC: National Aeronautics and Space Administration, 1980. Designed for school classes to familiarize them with the Shuttle and its mission. Well-illustrated and written in a catchy style, it is a good example of the public relations material put out by the agency.

Taylor, L.B., Jr. *Space Shuttle*. New York: Thomas Crowell, 1979. This book describes the reasons for and the design and operation of NASA's Space Shuttle and discusses who will be flying it and the benefits to be derived from its operation. It is a short work, and has been oriented toward a juvenile audience.

The Space Shuttle Adventure. Los Angeles, CA: Cheerios and Rockwell International, 1985. This is a short, 25-page booklet describing the Shuttle and its mission for young readers. It was put together as a promotional handout to capitalize on the popularity of the Shuttle.

Vogt, Gregory. *The Space Shuttle*. New York: Franklin Watts, 1983. This work discusses experiments proposed by high school students that have been performed aboard Skylab and gives advice to those interested in similar space research competitions for the Shuttle. It does include bibliographical references and an index. The book is definitely designed for a teenage audience.

Link to Part 2 (1992–2011), Chapter 14—Juvenile Literature about the Space Shuttle

Toward a History
of the
Space Shuttle

An Annotated Bibliography

Part 2, 1992–2011

National Aeronautics and
Space Administration

NASA

Monographs in Aerospace History, Number 49

TOWARD A HISTORY OF THE
SPACE SHUTTLE

AN ANNOTATED BIBLIOGRAPHY, PART 2 (1992–2011)

Compiled by
Malinda K. Goodrich
Alice R. Buchalter
Patrick M. Miller
of the
Federal Research Division, Library of Congress

NASA History Program Office
Office of Communications
NASA Headquarters
Washington, DC

Monographs in Aerospace History
Number 49
August 2012
NASA SP-2012-4549

PREFACE

This annotated bibliography is a continuation of *Toward a History of the Space Shuttle: An Annotated Bibliography*, compiled by Roger D. Launius and Aaron K. Gillette, and published by NASA as Monographs in Aerospace History, Number 1 in December 1992 (available online at *http://history.nasa.gov/Shuttlebib/contents.html*). The Launius/Gillette volume contains those works published between the early days of the United States' manned spaceflight program in the 1970s through 1991. The articles included in the first volume were judged to be most essential for researchers writing on the Space Shuttle's history. The current (second) volume is intended as a follow-on to the first volume. It includes key articles, books, hearings, and U.S. government publications published on the Shuttle between 1992 and the end of the Shuttle program in 2011. The material is arranged according to theme, including: general works, precursors to the Shuttle, the decision to build the Space Shuttle, its design and development, operations, and management of the Space Shuttle program. Other topics covered include: the *Challenger* and *Columbia* accidents, as well as the use of the Space Shuttle in building and servicing the Hubble Space Telescope and the International Space Station; science on the Space Shuttle; commercial and military uses of the Space Shuttle; and the Space Shuttle's role in international relations, including its use in connection with the Soviet *Mir* space station. This volume also includes juvenile literature on the Shuttle, as well as information about the Shuttle astronauts, memoirs about the Shuttle, and the end of the Space Shuttle program. A glossary of NASA abbreviations is included as well.

The contents of this bibliography were collected from a number of publications, including: *Ad Astra, Air & Space, Aviation Week & Space Technology, Economist, Interavia, Nature, New Scientist, New Yorker, New York Times, Newsweek, Popular Science, Science, Spaceflight, Space News, Space Policy, Technology and Culture, Time*, and *Washington Post*. Relevant publications of the Congressional Budget Office, congressional hearings, Congressional Research Service, and Government Accountability Office are included as well.

TABLE OF CONTENTS

NASA ABBREVIATIONS

AAS	American Astronomical Society
ANDE	Atmospheric Neutral Density Experiment
ARC	Ames Research Center
ASAP	Aerospace Safety Advisory Panel
ASTER	Advanced Spaceborne Thermal Emission and Reflection Radiometer
ATLAS	Atmospheric Laboratory for Applications and Science
CAIB	Columbia Accident Investigation Board
CBO	Congressional Budget Office
CEV	crew exploration vehicle
CLV	crew launch vehicle
COTS	Commercial Orbital Transportation System
CSA	Canadian Space Agency
CRS	Congressional Research Service
DFRC	Dryden Flight Research Center
DOD	U.S. Department of Defense
EDT	Eastern Daylight Time
ELC	Express Logistics Carrier
ELV	expendable launch vehicle
EPOCh	Extrasolar Planet Observations and Characterization
EPOXI	Extrasolar Planet Observations and Characterization (EPOCh) and Deep Impact Extended Investigation (DIXI)
ESA	European Space Agency
EVA	extravehicular activity
GAO	Government Accountability Office
GPS	global positioning system
GRACE	Gravity Recovery and Climate Experiment
GSFC	Goddard Space Flight Center
HST	Hubble Space Telescope
IML	International Microgravity Laboratory
IRVE	Inflatable Re-entry Vehicle Experiment
ISRO	Indian Space Research Organisation
ISS	International Space Station
JEF	Japanese Exposed Facility
JPL	Jet Propulsion Laboratory
JSC	Johnson Space Center
JWST	James Webb Space Telescope
KARI	Korea Aerospace Research Institute
KSC	Kennedy Space Center
KSLV	Korea Space Launch Vehicle
LaRC	Langley Research Center
LAS	launch-abort system
LIDAR	laser-imaging detection and ranging
LITE	Lidar In Space Technology Experiment
MISSE	Materials International Space Station Experiments

NASA ABBREVIATIONS

MLAS	Max Launch Abort System
MODIS	Moderate Resolution Imagine Spectroradiometer
MOU	memorandum of understanding
MRM	Mini Research Module
MSFC	Marshall Space Flight Center
MSL	Microgravity Science Laboratory
NAC	NASA Advisory Council
NAS	National Academy of Sciences
NASDA	National Space Development Agency of Japan
NCAR	National Center for Atmospheric Research
NESC	NASA's Engineering and Safety Center
NGLLXPC	Northrop Grumman Lunar Lander X Prize Challenge
NOAA	National Oceanic and Atmospheric Administration
NRC	National Research Council
NSF	National Science Foundation
OAST	Office of Aeronautics and Space Technology
OPM	Office of Personnel Management
PORT	Post-Landing Orion Recovery Test
RLV	reusable launch vehicle
RTF	Return to Flight
SMD	Science Mission Directorate
SRB	solid rocket booster
SRL	Space Radar Laboratory
SSC	Stennis Space Center
SSME	Space Shuttle main engine
SST	Spitzer Space Telescope
STS	Space Transportation System
STScI	Space Telescope Science Institute
TDRS	Tracking and Data Relay Satellites
Titan	Telemetry, Information, Transfer, and Attitude Navigation
USML	United States Microgravity Laboratory
USMP	U.S. Microgravity Payload
WFF	Wallops Flight Facility
WSF	Wake Shield Facility

CHAPTER 1—GENERAL WORKS

Asker, James R. "At 15, a Safer, Cheaper Shuttle." *Aviation Week & Space Technology*, 8 April 1996. This article examines the Space Shuttle program's status in 1996, the fifteenth anniversary of its creation, comparing it to the program as originally envisioned—as an inexpensive means of access to space. The author discusses how the program has changed over time, explaining how it has failed to meet its early goals.

Bergin, Mark. *Space Shuttle*. New York: Franklin Watts, 1999. This book, recommended for elementary and junior high school readers, follows the development of the Space Shuttle program from the early years to the present day, covering construction of the Shuttle and missions in space, including takeoff, reentry, and landing.

Bizony, Piers. *The Space Shuttle: Celebrating Thirty Years of NASA's First Space Plane*. Minneapolis: Zenith Press–MBI Publishing, 2011. This illustrated book marks a special moment in history: STS-134, the final mission of Space Shuttle *Endeavour*. The book provides a retrospective of all 134 Space Shuttle missions, including the Shuttle's final flight. In addition, the author discusses the development and design of the Space Shuttle, its technical specifications, and the details of its major assemblies and subassemblies.

Cowling, Tim. *The Space Shuttle*. Videocassette (VHS). Bethesda, MD: Program Enterprises, 1994. This video describes the tasks associated with the Space Shuttles' safe launch and return to Earth. Individuals around the world recount how they help make NASA and the Space Shuttle program work, describing jobs ranging from cleaning the Shuttle to intricate computer programming.

Dailey, J. R. "The First Space Shuttle." *Air & Space*, July 2004. This article examines the restoration of Space Shuttle *Enterprise* for exhibition at the Smithsonian Institution's Udvar-Hazy Center near Washington Dulles International Airport in Chantilly, Virginia, outside Washington, DC.

Drendel, Lou, and Ernesto Cumpian. *Walk Around Space Shuttle*. Carrollton, Texas: Squadron Signal Publications, 1999. This book provides a comprehensive collection of photographs of NASA Space Shuttles, including close-up photographic views of the Space Shuttles, with detailed captions.

Easterbrook, Gregg. "The Space Shuttle Must Be Stopped." *Time*, 25 July 2005. The author argues that the Space Shuttle program is too risky and too costly to maintain. He contends that Shuttle budgets use up funds that NASA could invest in developing a modern, safe, and less expensive spacecraft.

Exciting Simulations and Iceberg Interactive. *Space Shuttle Mission Simulator*. Space Simulation Computer Game. Haarlem, The Netherlands: Iceberg Interactive, 2010. This computer program enables users to experience the liftoff shakes and the roar of the engines, from the seat of the Space Shuttle commander or pilot. Users can perform simulated mission procedures, such as pressing buttons, rotating knobs, controlling the computer, and manually guiding the Shuttle to a safe landing.

Ferris, Timothy. "Earthbound." *New Yorker*, 1 August 1994. The author claims that the Space Shuttle program takes too much of NASA's budget and suggests that NASA should focus on exploring Mars.

Follett Software Company and Amazing Media. *Space Shuttle*. Cupertino, CA: KidSoft, 1996. DVD. This DVD program contains a multimedia introduction to the Space Shuttle program, including descriptions of space vehicles, equipment, and crews; orientation and training at NASA's Johnson Space Center; and details of living and working in space on 53 NASA missions.

Green, Barbara E. *United States Space Shuttle Firsts, 25th Anniversary*. KSC Historical Report 19, Library Archives, Kennedy Space Center, NASA, Florida, April 2006. The author compiled this summary of the Space Shuttle program from various reference publications available in the Library Archives of NASA's Kennedy Space Center. The report features photographs and brief information about significant events in the Space Shuttle program, from flights STS-1 to STS-114 (1981–2005). The report covers the Shuttle's first free flight in 1977; the first Shuttle launch in 1981; the first flight by an American woman astronaut, Sally K. Ride, in 1983; and the return to space of astronaut and U.S. Senator John H. Glenn Jr. in 1998.

Handberg, Roger, Joan Johnson-Freese, and George Moore. "The Myth of Presidential Attention to Space Policy." *Technology in Society* 17, no. 4 (27 December 1999): 337–348. The authors propose that, ever since President John F. Kennedy launched the race to place a U.S. astronaut on the Moon, the public has tended to regard the president as the prime mover behind the United States' civilian space program. The article argues that NASA's implicit goal throughout the years has been to recapture that original "magical" moment, suggesting that, besides working to make the space program economically viable, NASA must find ways to appeal to policymakers and the public, capturing their imagination and harnessing their support.

Harland, David M. *The Space Shuttle: Roles, Missions and Accomplishments*. New York: Wiley, 1998. This comprehensive book about the Space Shuttle describes its origins, operations, and explorations, as well as discussing weightlessness, exploration, and outposts. The book also contains a Space Shuttle mission log, glossary, and bibliography.

Hoversten, Paul. "The Truck: Satellites, Experiments, Space Station Parts—The Space Shuttle Hauled It All." *Air & Space*, August 2010. In this article discussing Space Shuttle payloads and the work carried out on the Shuttle, the author details the numerous objects the Shuttle has carried into space, including the Magellan Venus probe in 1989, the Galileo Jupiter probe in 1989, the HST, and pieces of the ISS.

Kay, W. D. "Democracy and Super Technologies: The Politics of the Space Shuttle and Space Station Freedom." *Science, Technology and Human Values* 19, no. 2 (April 1994): 131–151. Using the U.S. Space Shuttle program and the Space Station Freedom program as examples, the author argues that the problems of technology-based projects like those of the U.S. space program are deeply rooted in the American political process itself. The article explains that, in the United States, political requirements for obtaining approval and funding for large, expensive research and development projects create conditions that reduce the likelihood of the projects' technological success.

Klesius, Michael. "The Evolution of the Space Shuttle." *Air & Space*, July 2010. In this article, the author discusses the history of the Space Shuttle program, explores modifications to the design of the Space Shuttle considered throughout the program, and examines the errors leading to the loss of *Challenger* in January 1986 and *Columbia* in February 2003.

Launius, Roger D. "Assessing the Legacy of the Space Shuttle." *Space Policy* 22, no. 4 (November 2006): 226–234. This article describes the Space Shuttle as a constraint on other NASA space options, a flexible space-access vehicle, a platform for science, and a symbol of American technological prowess. The author argues that the Space Shuttle deserves a positive assessment because it helped foster development of spaceflight.

Lee, Wayne. *To Rise from Earth: An Easy-To-Understand Guide to Spaceflight.* 2nd ed. New York: Facts on File, 2000. The author explains the principles of rocket propulsion and discusses how the Space Shuttle achieves or changes orbit, describing the Shuttle's rockets and orbital mechanics. The book includes a chapter on space history, which traces the space milestones of the United States and Russia, and offers insight into the future of Mars exploration.

"Many Happy Returns? NASA at 50." *Economist*, 26 July 2008. The Space Shuttle and the ISS consume two-thirds of NASA's budget for piloted spaceflight. Claiming that the Space Shuttle program has been far more expensive than the throw-away rockets it was intended to replace, the author of this article argues that NASA does not justify the worth of its programs simply in terms of its balance sheet.

Morring, Frank, Jr. "Because It's Hard." *Aviation Week & Space Technology*, 29 September 2008. This article looks at issues facing NASA as it celebrates its 50th anniversary amid shifting national priorities in an ever-changing geopolitical scene.

"NASA Unveils Future Space Exploration Architecture." *Interavia Business & Technology*, Autumn 2005. This article unveils the architecture of NASA's future space exploration, explaining NASA's plan to use an improved, blunt-body capsule for its next generation of spacecraft. The new spacecraft will have a shorter development time, reduced reentry loads, increased landing stability, safe travel for up to six crew members, and the ability to dock at the ISS. The new system will have two primary launch vehicles—the crew-launch vehicle and the lunar heavy-cargo launch vehicle.

Official NASA Films Documenting the U.S. in Space. 5 vols. Videocassette (VHS). Burbank, CA: Warner Home Video, 1993. This five-volume videocassette collection of NASA films documents the U.S. space program. Two of the five volumes cover the Space Shuttle program. Volume 2, "Space Shuttle: From the Beginning," traces the history of the program, from the initial research and development of the design for a reusable spacecraft, through the first Shuttle missions. Volume 3, "Space Shuttle: Training, Facilities, Space Station," follows a Space Shuttle mission crew and vehicle from training to the day of the launch. The remaining volumes in the set cover other NASA programs.

"Old, Unsafe, and Costly." *Economist*, 30 August 2003. The article comments on proposals to discontinue the Space Shuttle program because of problems in Shuttle design, the costs of the program, and the safety risk of launching these space vehicles into space. In addition, the article discusses the possibility that the Space Shuttle program is having a negative effect on the development of a private space industry. The author believes that the Shuttle program has failed and that NASA should concentrate on developing high-risk technologies with the potential to transform routine space travel for people and equipment.

Pace, Scott. "Challenges to US Space Sustainability." *Space Policy* 25, no. 3 (August 2009): 156–159. The author describes space program sustainability issues over the coming decade, such as U.S. human access to space and the U.S. space industrial base, as well as long-term sustainability issues, such as protecting the space environment and the effects of space weather.

Pelton, Joseph N. "The Space Shuttle—Evaluating and American Icon." *Space Policy* 26, no. 4 (November 2010): 246–248. The author details key U.S. space policy decisions, as they relate to the Space Shuttle, and reviews the Space Shuttle program, including its origins in the White House of the President Richard M. Nixon administration. The article discusses the U.S. space transportation system, deployment of the ISS, the HST, and classified missions.

Pielke, Roger A., Jr. "Space Shuttle Value Open to Interpretation." *Aviation Week & Space Technology*, 26 July 1993. The article explains Space Shuttle costs from the point of view of a taxpayer, a space policymaker, and a national policymaker.

Review of U.S. Human Space Flight Plans Committee. "Seeking A Human Spaceflight Program Worthy of A Great Nation." Final Report, Washington, DC, October 2009. *http://www.nasa.gov/pdf/396093main_HSF_Cmte_FinalReport.pdf* (accessed 10 April 2012). The result of the White House Office of Science and Technology Policy's call for an independent review of the present and planned human spaceflight program, this report is also known as the "Augustine Report," after Norman R. Augustine, the chair of the committee that conducted the review. The White House–appointed Review of U.S. Human Space Flight Plans Committee comprised 10 members with diverse professional backgrounds, including scientists, engineers, astronauts, educators, executives of established and new aerospace firms, former presidential appointees, and a retired Air Force General. This report describes the independent review of the current program and suggests alternatives that would ensure that the nation is pursuing a safe, innovative, affordable, and sustainable trajectory for the future of human spaceflight.

Rumerman, Judith A., Chris Gamble, and Gabriel Okolski. "Space Shuttle." In *U.S. Human Spaceflight: A Record of Achievement, 1961–2006*, 35–74. Monographs in Aerospace History, no. 41. NASA Report no. SP-4541, NASA History Division, Office of External Relations, NASA Headquarters, Washington, DC, December 2007. *http://history.nasa. gov/monograph41.pdf* (accessed 8 March 2012). This chapter of NASA's book on human spaceflight describes Space Shuttle flights STS-1 through STS-116 (1981–2006) and provides information about each Shuttle mission, including crew-member names, Shuttle names, and the purpose of each mission. The chapter also provides a short bibliography on the Shuttle.

Smith, Marcia S. "Space Exploration: Issues Concerning the 'Vision for Space Exploration'." CRS Report for Congress, Congressional Research Service, Library of Congress, Washington, DC, 6 September 2005. *http://assets.opencrs.com/rpts/RS21720_ 20050609.pdf* (accessed 8 March 2012). This report provides an overview of President George W. Bush's 2004 Vision for Space Exploration and congressional reaction to the Vision, which includes terminating the Space Shuttle program in 2010.

Space Shuttle: A Remarkable Flying Machine. Videocassette (VHS). Houston, Texas: TaLas Enterprises, 1994. This film documents the first historic flight of a Space Shuttle, the U.S. spacecraft *Columbia*, which launched on 12 April 1981. The footage highlights liftoff, the on-board activities of astronauts John W. Young and Robert L. Crippen, and the Shuttle's landing in Rogers Dry Lake bed in California. This file is available for download from: *http://www.archive.org/details/gov.archives.arc.1157922*.

The History of the Space Shuttle. Atlanta, GA: Whitman Publishing, 2012. *http://whitman.com/ Inventory/Detail/The-History-of-the-Space-Shuttle* (accessed 10 April 2012). This book recounts the history of NASA's Space Shuttle program, from earliest design and testing of the Shuttle to its historic final launch on 8 July 2011. Each chapter includes links to more than 70 audio and video clips explaining the critical role that the Shuttle has played in establishing a human presence in low Earth orbit, as well as describing Shuttle missions and crews and the experience of eating, sleeping, and living aboard the spacecraft.

The Space Shuttle. DVD. New York: A & E Television Networks and History Channel Club, 2008. This DVD program, containing footage from missions, expert interviews, and computer simulations, describes the Space Shuttle from its conception to its launch, recounting the story of how NASA overcame risks and challenges to make the Space Shuttle a reality.

The Space Shuttle. DVD. New York: Jaffe Productions, Hearst Entertainment, A & E Home Video, New Video Group, and History Channel, 2004. This DVD program details the development of the Space Shuttle from the 1950s to its triumphant launch in 1981. The program examines the successes and failures of the Shuttle missions and looks at the next generation Shuttle—the futuristic commercial reusable space vehicle X-33 VentureStar.

Tsuruta, Nobuyuki, and Robert Lanchester. *The Space Shuttle*. DVD. Princeton, NJ: Films for the Humanities and Sciences, 2003. This DVD program, an analysis of the Space Shuttle, explains Shuttle technology and shows the functions of the three components of the Space Shuttle system: the orbiter, the external tank, and the solid rocket boosters. The film demonstrates how the orbiter's flight deck functions and explains the operation of its attitude control and orbital maneuvering systems, as well as discussing its payload bay equipped with a Remote Manipulation System.

Vedda, James A. "Space Power." *Ad Astra*, February 2003. This article focuses on the U.S. Congress's behavior toward NASA's human spaceflight programs and the Space Shuttle since the end of the Apollo program, examining how institutional changes in Congress over the previous 30 years could affect policymaking for current and future civil space activities.

Link to Part 1 (1970–1991), Chapter 1—General Works

CHAPTER 2—PRECURSORS

Heppenheimer, T. A. *Countdown: A History of Space Flight*. New York: Wiley, 1999. Taking a historian's view of the space age, the author places the U.S. space program's major achievements, such as the Apollo program, in the context of concurrent political and social developments. The author juxtaposes the space programs of the U.S.S.R. and the United States, providing insights into why the two most powerful nations in the world became embroiled in the costly and politically charged race for space.

Heppenheimer, T. A. *Space Shuttle Decision, 1965–1972*. 2 vols. Washington, DC: Smithsonian Institution Press, 2002. The author looks back at the Space Shuttle's technical antecedents, such as the X-15 rocket plane and rocket-booster technologies. In addition, he illuminates the principal personalities involved in the decision to build the Space Shuttle and considers their motivations. Tracing the development of the myriad designs that preceded the Shuttle concept, he discusses NASA's evolving technical calculations, its fiscal constraints, and the background of political maneuvering that influenced the development of program goals.

Jenkins, Dennis R. *Space Shuttle: The History of Developing the National Space Transportation System, the Beginning Through STS-75*. Marceline, MO: Walsworth Publishing Company, 1997. This book discusses technical approaches to building an RLV or a semi-RLV. In addition, the author explains the history of the Space Shuttle, including its funding, construction, and flights. The book also describes the technology of the Space Shuttle, as it was originally built and as it has been modified through the years.

Launius, Roger D., and Dennis R. Jenkins, eds. *To Reach the High Frontier: A History of U.S. Launch Vehicles*. Lexington, KY: University Press of Kentucky, 2002. Because of the technical challenge of reaching space with chemical rockets, the high costs associated with space launch, the long lead times necessary for scheduling flights, and the poor reliability of the rockets themselves, launch vehicles are the space program's most difficult challenge. This book recounts the history of each type of space-access vehicle developed in the United States since the birth of the space age in 1957. Two chapters cover the Space Shuttle.

Overbye, Dennis. "As Shuttle Era Ends, Dreams of Space Linger." *New York Times*, 5 July 2011. Written on the eve of the final Space Shuttle flight, STS-135, Space Shuttle *Atlantis*'s last mission before retirement, this article reviews the programs that led to development of the Shuttle.

Reed, Dale R., and Darlene Lister. *Wingless Flight: The Lifting Body Story*. Report no. NASA SP-4220, NASA History Division, Office of Policy and Plans, Washington, DC, 1996. *http://history.nasa.gov/SP-4220/sp4220.htm* (accessed 8 March 2012). This book examines the lifting body, a wingless vehicle that was a precursor to the Space Shuttle. This spacecraft flew because of the lift generated by the shape of its fuselage. One of these lifting bodies, the X-24B, shared very similar speed and performance characteristics with the projected Shuttle spacecraft design. The X-24B was used to collect operational data used in the design and development of the Space Shuttle vehicles.

Rumerman, Judith A., Chris Gamble, and Gabriel Okolski. *U.S. Human Spaceflight: A Record of Achievement, 1961–2006*, 35–74. Monographs in Aerospace History, no. 41. NASA Report no. SP-4541, NASA History Division, Office of External Relations, NASA Headquarters, Washington, DC, December 2007. *http://history.nasa.gov/monograph41. pdf* (accessed 8 March 2012). This reference book contains information on the NASA manned space exploration programs, including the Mercury, Gemini, Apollo, Skylab, and Apollo-Soyuz programs. The book also includes chapters on the Space Shuttle and International Space Station programs. The Space Shuttle chapter describes the program and provides brief descriptions of the first 117 Shuttle flights. The book also contains many black and white photographs and an appendix of Shuttle main payloads.

The Space Shuttle. DVD. New York: A & E Television Networks and History Channel Club, 2008. This DVD program, containing footage from missions, expert interviews, and computer simulations, describes the Space Shuttle from its conception to its launch, recounting the story of how NASA overcame risks and challenges to make the Space Shuttle a reality.

The Space Shuttle. DVD. New York: Jaffe Productions, Hearst Entertainment, A & E Home Video, New Video Group, and History Channel, 2004. This DVD program details the development of the Space Shuttle from the 1950s to its triumphant launch in 1981. The program examines the successes and failures of the Shuttle missions and looks at the next generation Shuttle—the futuristic commercial reusable space vehicle X-33 VentureStar.

Thompson, Milton O. *Flight Research: Problems Encountered and What They Should Teach Us.* With a Background Section by J. D. Hunley. Monographs in Aerospace History no. 22, NASA Report no. SP-2000-4522, NASA History Division, Office of Policy and Plans, NASA Headquarters, Washington, DC, 2000. *http://www.nasa.gov/centers/dryden/pdf/ 88795main_Thompson.pdf* (accessed 8 March 2012). This book describes early aeronautic research that helped lay the groundwork for the development of the Space Shuttle. The author discusses aerodynamics, environmental systems, control systems, landing gear, and heating issues.

Thompson, Milton O., and Curtis Peebles. *Flying Without Wings: NASA Lifting Bodies and the Birth of the Space Shuttle*. Washington, DC: Smithsonian Institution, 1999. This book charts the transformation of aircraft into spacecraft, describing the efforts of a small group of NASA pilots and researchers to prove a seemingly impossible aerodynamic concept that would profoundly influence the history of spaceflight. The authors explain how, after the cancellation of the U.S. Air Force's Dyna-Soar Program, the first lifting body projects, such as the Parasev paraglider and the M2-F1, were built on shoestring budgets at Edwards Air Force Base, California, often without the knowledge of officials at the NASA Headquarters.

Wallace, Lane E. "Flights of Discovery: 50 Years at the NASA Dryden Flight Research Center." NASA Report no. SP-4309, NASA History Series, Washington, DC, 1996. This history of the first 50 years at NASA's Dryden Flight Research Center captures the spirit of aeronautical research and development from 1946 to 1996, providing insightful accounts of most of the major flight research projects during this period. NASA used Dryden Flight Research Center to engineer and test the Space Shuttle and its precursors.

Link to Part 1 (1970–1991), Chapter 2—Precursors of the Shuttle

CHAPTER 3—THE DECISION TO BUILD THE SPACE SHUTTLE

Covault, Craig. "Blame It on Nixon; Space Policy Failures Bred NASA Shuttle Promises That Were Unattainable." *Aviation Week & Space Technology*, 19 March 2007. In this article, Covault discusses the effect on the U.S. space program of President Richard M. Nixon's 1972 approval of the plan to develop the Space Shuttle. In March 1973, James C. Fletcher of NASA announced that further studies had uncovered additional potential Shuttle uses, and that NASA expected the new uses would increase the program's cost benefits.

Gold, Thomas. "Is NASA an Expensive Failure?" *Nature* 366, no. 6457 (23 December 1993): 723. This opinion article decries NASA's insecure, publicity-seeking policies, which have made a debacle of the U.S. space program. The author notes that, although NASA originally claimed that Space Shuttle space launches would be much cheaper than disposable rocket launches, Shuttle launch costs have actually been much higher than those of unpiloted, disposable boosters. Higher Space Shuttle costs have meant the loss of other NASA projects, such as Mars Orbiter.

Heppenheimer, T. A. *Space Shuttle Decision, 1965–1972*. Vol. 1 of *History of the Space Shuttle*. The NASA History Series. Washington, DC: Smithsonian Institution Press, 2002. The author looks back at the Space Shuttle's technical antecedents, such as the X-15 rocket plane and rocket-booster technologies. In addition, he illuminates the principal personalities involved in the decision to build the Space Shuttle and considers their motivations.

Jenkins, Dennis R. *Space Shuttle: The History of Developing the National Space Transportation System, the Beginning Through STS-75*. Marceline, MO: Walsworth Publishing Company, 1997. This book discusses technical approaches to building an RLV or a semi-RLV. In addition, the author explains the history of the Space Shuttle, including its funding, construction, and flights. The book also describes the technology of the Space Shuttle, as it was originally built and as it has been modified through the years.

Klesius, Michael. "The Evolution of the Space Shuttle." *Air & Space*, July 2010. In this article, the author discusses the history of the Space Shuttle program, explores modifications to the design of the Space Shuttle considered throughout the program, and examines the errors leading to the loss of *Challenger* in January 1986 and *Columbia* in February 2003.

Launius, Roger D. "History Reference Center: NASA and the Decision To Build the Space Shuttle, 1969–72." *Historian* 57, no. 1 (Autumn 1994): 17. This article describes NASA's efforts between 1969 and 1972 to create a viable, low-cost Space Shuttle program. Other topics covered include completion of the Apollo mission, NASA's participation in a political decision-making process, and President Richard M. Nixon's appointment of the Space Task Group to study possible future projects after the end of the Apollo program.

Temple, L. Parker, III. "Committing to the Shuttle Without Ever Having a National Policy." *Air Power History*, 22 September 2005. This article claims that President Richard M. Nixon's original 1969 U.S. space policy never made an explicit commitment to the use of the Space Shuttle for all U.S. launches. The author seeks to understand how the de facto commitment developed, arguing that this policy has nearly led to the United States' sacrificing its ability to produce expendable launch vehicles. He concludes that a finesse of the policy process was responsible for this outcome.

U.S. General Accounting Office. "1998 NASA Budget: Review of Selected Activities." Report no. GAO/NSIAD-97-252R, Washington, DC, 30 September 1997. *http://archive.gao.gov/ paprpdf1/159356.pdf* (accessed 8 March 2012). This report reviews NASA's FY 1998 budget request. The GAO notes that it has identified opportunities to reduce NASA's FY 1998 budget request by about US$108 million, primarily involving cutbacks to human spaceflight programs (US$54.4 million) and mission support (US$53 million).

Vedda, James A. "Evolution of Executive Branch Space Policy." *Space Policy* 12, no. 3 (August 1996): 177–192. This article describes space policies enacted by different presidents and explains how these policies affected Space Shuttle development. NASA had originally conceived of the Space Shuttle as a delivery vehicle, but in response to presidential policy changes, NASA developed a Space Shuttle that could function as a laboratory and show economic returns.

Link to Part 1 (1970–1991), Chapter 3—The Shuttle Decision

See also Part 1 (1970–1991), Chapter 8—Shuttle Promotion

CHAPTER 4—DESIGN AND DEVELOPMENT

Biggs, R. E. *Space Shuttle Main Engine: The First Twenty Years and Beyond.* American Astronautical Society History Series, vol. 29. San Diego, CA: Univelt, 2008. This article describes the Space Shuttle's main engine design and development, outlining requirements, obstacles, component testing, and engine testing. Discussing the *Challenger* tragedy and the Space Shuttle's return to flight, the author recounts NASA's efforts to attain full-power-level certification and acceptance tests for the Shuttle.

Bizony, Piers. *The Space Shuttle: Celebrating Thirty Years of NASA's First Space Plane.* Minneapolis: Zenith Press–MBI Publishing, 2011. This illustrated book marks a special moment in history: STS-134, the final mission of Space Shuttle *Endeavour.* The author discusses the development and design of the Space Shuttle, its technical specifications, and the details of its major assemblies and subassemblies.

Blomberg, Richard D. "Report on Shuttle Safety." *Ad Astra*, August 2002. This article summarizes the Space Shuttle safety issues addressed by NASA's Aerospace Safety Advisory Panel (ASAP), along with the panel's findings and recommendations regarding Space Shuttle plans and budgetary requests. The author also makes recommendations regarding ground infrastructure and launch workforce.

Chien, Philip. "Shuttle Gets a Boost." *Popular Science*, August 1999. This article describes the avionics upgrade for Space Shuttle *Atlantis.*

Corneille, Philip. "Refurbishing the Shuttle's SRBs." *Spaceflight*, August 2001. This article describes how Space Shuttle solid rocket boosters (SRBs) are made, launched, and recovered, as well as SRB refurbishment and reassembly. The author explains how SRBs are refurbished for use on later Shuttle missions.

Covault, Craig. "Roaring Comeback." *Aviation Week & Space Technology*, 11 July 2005. This article discusses the Space Shuttle main engines (SSMEs) and solid rocket boosters (SRBs) that will propel *Discovery*'s launch into orbit in its return-to-flight mission. In the two years since the *Columbia* accident, NASA's SSME and SRB programs have substantially increased the rigor of their testing and quality oversight.

Covault, Craig, and Frank Morring Jr. "Re-Discovery; As Discovery Roars Back to Space, Its Flight-Test Lessons Warn That the Program Still Needs Critical Work Back Here." *Aviation Week & Space Technology*, 1 August 2005. During STS-114 in 2005, *Discovery* sustained a low-speed bird strike at liftoff, which caused insulation to separate from the external tank and a piece of foam to come off the liquid hydrogen tank interface ring. The incidents reinforced the need for Marshall and Lockheed Martin to perform additional critical materials and manufacturing process work at its Michoud, Louisiana, tank assembly plant.

Heppenheimer, T. A. *Development of the Space Shuttle, 1972–1981*. Vol. 2 of *History of the Space Shuttle*. The NASA History Series. Washington, DC: Smithsonian Books, 2010. This book focuses on the engineering challenges of building the Space Shuttle: the development of propulsion, thermal protection, electronics, and on-board systems. The author traces the Shuttle's development through the decade of engineering setbacks and breakthroughs, program-management challenges, and political strategizing that culminated in the first launch of the Space Shuttle in April 1981. He also discusses the planning and preparation for the first Shuttle launch.

Hunley, J. D. *U.S. Space-Launch Vehicle Technology: Viking to Space Shuttle*. Gainesville, FL: University Press of Florida, 2008. This book examines hardware designed explicitly for launching vehicles into space and traces the evolution of that technology from the early Vanguard rocket program through the Space Shuttle. The author describes propulsion systems, structural issues, and guidance-and-control devices. He concludes that the major factors advancing the development of this technology included high levels of Cold War funding, an engineering culture that promoted technology transfer and was flexible enough to overcome unexpected problems, and management systems that furnished systems engineering and cost control over a huge number of organizations.

Inside the Space Shuttle. Videocassette (VHS). Port Washington, NY: Koch Vision, 1998. This video provides a behind-the-scenes glimpse into NASA's world of high technology, showing the most amazing flying machines humans have ever created—the Space Shuttles.

Jenkins, Dennis R. *Space Shuttle: The History of Developing the National Space Transportation System, the Beginning Through STS-75*. Marceline, MO: Walsworth Publishing Company, 1997. This book discusses technical approaches to building an RLV or a semi-RLV. In addition, the author explains the history of the Space Shuttle, including its funding, construction, and flights. The book also describes the technology of the Space Shuttle, as it was originally built and as it has been modified through the years.

Klesius, Michael. "The Evolution of the Space Shuttle." *Air & Space*, July 2010. In this article, the author discusses the history of the Space Shuttle program, explores modifications to the design of the Space Shuttle considered throughout the program, and examines the errors leading to the loss of *Challenger* in January 1986 and *Columbia* in February 2003.

Lee, Wayne. *To Rise from Earth: An Easy-To-Understand Guide to Spaceflight*. 2nd ed. New York: Facts on File, 2000. The author explains the principles of rocket propulsion and discusses how the Space Shuttle achieves or changes orbit, describing the Shuttle's rockets and orbital mechanics. The book includes a chapter on space history, which traces the space milestones of the United States and Russia, and offers insight into the future of Mars exploration.

Oberg, James. "Puncture Repair Kit; The Shuttle Columbia Might Have Been Saved If the Crew Had Been Able To Fix A Hole While in Orbit." *New Scientist*, 15 November 2003. This article examines NASA's efforts to create patch kits that astronauts could use to make repairs to orbiters if they sustained damage such as *Columbia* experienced during its last takeoff. Besides trying out NASA's repair kit, the author identifies other options for inflight repairs to external surfaces of the spacecraft.

U.S. General Accounting Office. "Space Shuttle: Declining Budget and Tight Schedule Could Jeopardize Space Station Support." Report no. GAO/NSIAD-95-171, Washington, DC, 28 July 1995. *http://www.gao.gov/archive/1995/ns95171.pdf* (accessed 8 March 2012). This report reviews NASA's efforts to redesign the lift capability of the Space Shuttle so that it will be able to make the 21 flights necessary to complete the assembly of the ISS within five years. The GAO found that NASA's plans for increasing the Shuttle's lift capability are complex, involving approximately 30 individual actions, such as hardware redesigns, improved flight-design techniques, and new operational procedures.

U.S. General Accounting Office. "Space Shuttle: Status of Advanced Solid Rocket Motor Program." Report no. GAO/NSIAD-93-26, Washington, DC, November 1992. *http://archive.gao.gov/d36t11/148147.pdf* (accessed 8 March 2012). This report describes NASA's Advanced Solid Rocket Motor Program, which is intended to increase the lift capacity of the Space Shuttle. The GAO expresses concerns that the motor's actual price has far exceeded the estimate and that NASA may not need the solid rocket motor after all.

U.S. General Accounting Office. "Space Shuttle Main Engine: NASA Has Not Evaluated the Alternate Fuel Turbopump Costs and Benefits." Report no. GAO/NSIAD-94-54, Washington, DC, October 1993. *http://archive.gao.gov/t2pbat5/150179.pdf* (accessed 8 March 2012). This report focuses on NASA's decision to develop an alternative high-pressure fuel turbopump for the Space Shuttle's main engine. The report recommends that, before deciding whether to resume development of the alternate fuel pump, the NASA Administrator should require NASA officials to estimate the life-cycle costs and benefits for the alternative fuel-pump program and to compare them with the costs and benefits of further improvements to the existing pump.

Van de Haar, Gerard, and Luc van den Ableen. "KSC's Operations and Checkout Building." *Spaceflight*, July 1995. This article describes the building at NASA's Kennedy Space Center where NASA engineers prepare Space Shuttles for launch and inspect them after they return from a mission. The authors describe both the capabilities of the facility and the activities that take place inside it.

Link to Part 1 (1970–1991), Chapter 4—Shuttle Design and Development

Link to Part 1 (1970–1991), Chapter 5—Space Shuttle Testing and Evaluation

CHAPTER 5—OPERATIONS

Anselmo, Joseph C. "NASA To Seek Major Shift in U.S. Shuttle Policy." *Aviation Week & Space Technology*, 13 October 1997. This article describes NASA's support for United Space Alliance's lobby to rescind an 11-year-old presidential edict and a provision in U.S. law prohibiting the Space Shuttle from carrying commercial satellites into orbit. United Space Alliance, the venture that manages Shuttle operations, is promoting these policy changes in the hope of completely privatizing Shuttle operations within five years. The expendable launch vehicle industry is likely to oppose the move so that it can retain its monopoly on commercial spacecraft launches.

Behar, Michael. "The Ground." *Air & Space*, November 2006. The article focuses on the responsibilities of flight directors in space explorations. To be certified to work an actual launch and landing of a Space Shuttle, a flight director must undergo hundreds of Shuttle simulations and be prepared to overcome disaster scenarios.

Brandon-Cremer, Lee, and Joel Powell. *Space Shuttle Almanac*. Calgary, AB: Microgravity Productions, 1992. *http://www.amazon.com/Space-Shuttle-Almanac-ebook/dp/ B005IWAVOA/ref=sr_1_8?s=books&ie=UTF8&qid=1323720909&sr=1-8* (accessed 8 March 2012). In this final digital edition of the *Space Shuttle Almanac*, primary author Lee Brandon-Cremer celebrates 40 years of Shuttle operational history within a 1,400-page compilation of mission facts, figures, dates, and times. The almanac includes an outstanding collection of more than 1,000 photographs and more than 1,000 diagrams, covering every mission. The e-book is available on CD or by download.

Evans, Ben. *Space Shuttle Columbia: Her Missions and Crews*. Chichester, UK: Praxis Publishing, 2005. This book, written by the scientists and researchers who developed and supported *Columbia*'s many payloads, the engineers who worked on the spacecraft, and the astronauts who flew it, comprises detailed descriptions of Space Shuttle *Columbia*'s 28 missions. The book is intended as a tribute to *Columbia* and to the people who have supported the Shuttle program.

Green, Andrew. "STS-125: Hubble's Final Refit in Orbit." *Spaceflight*, August 2009. This article describes STS-125, Space Shuttle *Atlantis*'s mission to repair and upgrade the HST. The article describes prelaunch activities, crew members, and daily activities during the 12-day mission.

Hoversten, Paul. "The Truck: Satellites, Experiments, Space Station Parts—The Space Shuttle Hauled It All." *Air & Space*, August 2010. In this article discussing Space Shuttle payloads and the work carried out on the Shuttle, the author details the numerous objects the Shuttle has carried into space, including the Magellan Venus probe in 1989, the Galileo Jupiter probe in 1989, the HST, and pieces of the ISS.

Kidger, Neville. "STS-119: Orbital Operations." *Spaceflight*, May 2009. This article describes activities aboard the ISS, as well as summarizing STS-119, Space Shuttle *Discovery*'s mission to deliver solar arrays to the ISS. The article lists the Shuttle crew members and describes the crew's activities during the 12-day mission.

Kidger, Neville. "STS-120: Orbital Operations." *Spaceflight*, January 2008. This article describes activities aboard the ISS, as well as summarizing STS-120, Space Shuttle *Discovery*'s mission to deliver Harmony Node, a connecting module that will increase the orbiting laboratory's interior space. STS-120 crew also repaired the ISS's solar arrays. The article lists the STS-120 crew members and describes day-to-day operations during the 16-day mission.

Kidger, Neville. "STS-122: Orbital Operations." *Spaceflight*, April 2008. This article describes activities aboard the ISS, as well as summarizing STS-122, Space Shuttle *Atlantis*'s mission to deliver and to install the Columbus laboratory, the ISS module contributed by ESA. The article lists STS-122 crew members and describes the crew's activities during the 13-day mission.

Kidger, Neville. "STS-123: Orbital Operations." *Spaceflight*, May 2008. This article describes activities aboard the ISS, as well as summarizing STS-123, Space Shuttle *Endeavour*'s mission to equip and supply the ISS. *Endeavour* delivered the Japanese Experiment Logistics Module, which contains avionics and will serve as a storage area for experiment materials. The article, which names STS-123 crew members and describes day-to-day operations during the 15-day mission, continues in the June 2008 edition of *Spaceflight*.

Kidger, Neville. "STS-123: Orbital Operations." *Spaceflight*, June 2008. This continues an article in the May 2008 edition of *Spaceflight* summarizing STS-123, Space Shuttle *Endeavour*'s mission to equip and supply the ISS. *Endeavour* delivered the Japanese Experiment Logistics Module, which contains avionics and will serve as a storage area for experiment materials. The article names STS-123 crew members and describes day-to-day operations during the 15-day mission.

Kidger, Neville. "STS-127: Orbital Operations." *Spaceflight*, October 2009. This article describes activities aboard the ISS, as well as summarizing STS-127, Space Shuttle *Endeavour*'s mission to deliver the Kibo Japanese Experiment Module Exposed Facility and Experiment Logistics Module Exposed Section. The article lists the Shuttle crew members and describes the crew's activities during the 15-day mission.

Kidger, Neville. "STS-128: Orbital Operations." *Spaceflight*, November 2009. This article describes activities aboard the ISS, as well as summarizing STS-128, Space Shuttle *Discovery*'s mission to deliver 7 tons (6.4 tonnes) of equipment and supplies to the ISS. After this mission, NASA will begin a transition from Shuttle missions to help build the ISS, to missions to use the ISS. The article lists crew members, recounts prelaunch activities, and describes activities during the 13-day mission.

Kremer, Ken. "STS-129: Shuttle Delivers Spares to ISS." *Spaceflight*, February 2010. This article describes STS-129, the mission of Space Shuttle *Atlantis* to equip and supply the ISS. The article lists crew members and daily activities during the 10-day mission.

Kremer, Ken. "STS-130: New Window on the World." *Spaceflight*, April 2010. This article describes STS-130, Space Shuttle *Endeavour*'s mission to deliver equipment and supplies to the ISS. The article describes prelaunch activities, crew members, and activities during the 13-day mission.

Kremer, Ken. "STS-131: Discovery's Penultimate Voyage." *Spaceflight*, June 2010. This article describes STS-131, in which Space Shuttle *Discovery* delivered 8 tons (7.3 tonnes) of equipment and supplies to the ISS. The article describes prelaunch activities, crew members, and day-to-day operations during the 15-day mission.

Kremer, Ken. "STS-132: Atlantis' Last Blast with Russian Beauty." *Spaceflight*, August 2010. This article describes STS-132, Space Shuttle *Atlantis*'s final mission before its retirement. The goal of STS-132 was to equip and supply the ISS. Besides carrying spare parts to the ISS, *Atlantis* delivered the 11,000-pound (4.989.5-kilogram) Russian Rassvet Mini Research Module, which will enable ISS crew to conduct biotechnology and fluid physics experiments. The article describes prelaunch activities, crew members, payloads, and daily operations during the 11-day mission.

Legler, Robert D., and Floyd V. Bennett. *Space Shuttle Missions Summary*. Hanover, MD: NASA Center for AeroSpace Information, September 2011. *http://www.scribd.com/ doc/70899668/Space-Shuttle-Missions-Summary* (accessed 8 March 2012). This is a handy reference guide for data on all Space Shuttle missions. The book provides "as-flown" data for ascent, on-orbit events, and descent mission phases, compiled from many flight-support sources. In addition, the book identifies the specific Shuttle vehicle configuration, payload, flight crew, and flight directors for each flight and includes pertinent photos on each mission summary page.

Marino, Joe. "Twenty Years of the Space Shuttle." *Ad Astra*, April 2001. This article presents photographs of Space Shuttle launches, including the liftoff of *Discovery, STS-64*, in September 1994.

NASA. Johnson Space Center. "History of Space Shuttle Rendezvous." Report no. JSC-63400 Revision 3, Mission Operations Directorate, Flight Dynamics Division, Johnson Space Center, NASA, Houston, TX, October 2011. *http://ntrs.nasa.gov/archive/nasa/ casi.ntrs.nasa.gov/20110023479_2011024697.pdf* (accessed 8 March 2012). This technical report provides an overview of rendezvous and proximity operations missions flown by the Space Shuttle from 1983 to 2011. Rendezvous profile evolution from 1969 to 2011 is covered, along with unique challenges faced by the Space Shuttle that were not encountered during the Gemini, Apollo, Skylab, and Apollo/Soyuz missions. Four chapters focus on the following: 1) STS-39 deploy/retrieve mission, 2) HST repair missions, 3) STS-130 mission to the ISS, and 4) the level of automation in U.S. spacecraft.

Powell, Joel W., and Lee Robert Caldwell. *The Space Shuttle Almanac: A Comprehensive Overview of the First Ten Years of Space Shuttle Operations*. Calgary, Alberta: Microgravity Productions, 1992. This reference tool provides an overview of Space

Shuttle operations, facilities, hardware, and missions for the first 39 Shuttle flights. Using information from NASA publications, the authors describe the Shuttle orbiter, discuss each mission flown, and provide details on crew, experiments, and payloads.

Rumerman, Judith A., Chris Gamble, and Gabriel Okolski. "Space Shuttle." In *U.S. Human Spaceflight: A Record of Achievement, 1961–2006*, 35–74. Monographs in Aerospace History, no. 41. NASA Report no. SP-4541, NASA History Division, Office of External Relations, NASA Headquarters, Washington, DC, December 2007. *http://history.nasa.gov/monograph41.pdf* (accessed 8 March 2012). This chapter of NASA's book on human spaceflight describes Space Shuttle flights STS-1 through STS-116 (1981–2006) and provides information about each Shuttle mission, including crew-member names, Shuttle names, and the purpose of each mission. The chapter also provides a short bibliography on the Shuttle.

Rumerman, Judy A., and Stephen J. Garber. *Chronology of Space Shuttle Flights, 1981–2000.* NASA Report no. HHR-70, NASA History Division, Office of Policy and Plans, NASA Headquarters, Washington, DC, October 2000. This book, prepared by NASA's History Office, provides a chronology of Space Shuttle flights from 1981 through 2000.

Schuiling, Roelof. "STS-45 Mission Report." *Spaceflight*, October 1992. This article discusses STS-45, the mission of Space Shuttle *Atlantis* to carry the first Atmospheric Laboratory for Applications and Science (ATLAS-1) into orbit. ATLAS-1 was designed to conduct studies in atmospheric chemistry, solar radiation, space plasma physics, and ultraviolet astronomy. The author describes crew members, daily activities, launch preparation, and payloads during the nine-day mission.

Schuiling, Roelof. "STS-46 Mission Report." *Spaceflight*, January 1993. This article discusses STS-46, the mission of Space Shuttle *Atlantis* to deploy ESA's European Retrievable Carrier and to operate the joint project of NASA and Agenzia Spaziale Italiana (the Italian space agency)—the Tethered Satellite System. The author describes launch preparation, payloads, crew members, and activities during the eight-day mission.

Schuiling, Roelof. "STS-47 Mission Report." *Spaceflight*, April 1993. This article discusses STS-47, Space Shuttle *Endeavour*'s mission to carry Spacelab-J into orbit and to use it to conduct experiments in the Shuttle payload. A joint mission of NASA and Japan's space agency, NASDA, Spacelab-J uses an unpiloted Spacelab module to conduct microgravity investigations in materials and life sciences. The author describes launch preparation, payloads, crew members, and daily activities during the eight-day mission.

Schuiling, Roelof. "STS-49 Mission Report." *Spaceflight*, August 1993. This article discusses STS-49, Space Shuttle *Endeavour*'s mission to capture the INTELSAT-VI (F-3) satellite, which was stranded in an unusable orbit, and to install a new perigee kick motor on the satellite. The author describes crew members, daily activities, launch preparation, and payloads during the 11-day mission.

Schuiling, Roelof. "STS-50 Mission Report." *Spaceflight*, November 1992. This article discusses STS-50, Space Shuttle *Columbia*'s mission to carry the United States Microgravity Laboratory 1 (USML-1), a piloted Spacelab module with a tunnel connecting to the crew compartment. USML-1 was designed to conduct containerless processing experiments, materials processing studies, and other experiments. The author describes crew members, daily activities, launch preparation, and payloads during the 14-day mission.

Schuiling, Roelof. "STS-51 Mission Report." *Spaceflight*, December 1993. This article discusses STS-51, Space Shuttle *Discovery*'s mission to deploy the Advanced Communications Technology Satellite and the Orbiting and Retrievable Far and Extreme Ultraviolet Spectrograph-Shuttle Pallet Satellite. The author describes crew members, daily activities, launch preparation, and payloads during the 11-day mission.

Schuiling, Roelof. "STS-52 Mission Report." *Spaceflight*, February 1993. This article discusses STS-52, Space Shuttle *Columbia*'s mission to deploy the Laser Geodynamic Satellite-2 and operate the U.S. Microgravity Payload-1. The author describes crew members, daily activities, launch preparation, and payloads during the 10-day mission.

Schuiling, Roelof. "STS-53 Mission Report." *Spaceflight*, March 1993. This article discusses STS-53, Space Shuttle *Discovery*'s mission to deploy a classified DOD satellite and to conduct two unclassified secondary experiments in the Shuttle payload bay. The two secondary experiments were the Orbital Debris Radar Calibration Spheres and the Shuttle Glow Experiment/Cryogenic Heat Pipe Experiment. The author describes crew members, daily activities, launch preparation, and payloads during the eight-day mission.

Schuiling, Roelof. "STS-54 Mission Report." *Spaceflight*, March 1993. This article discusses STS-54, Space Shuttle *Endeavour*'s mission to deploy the fifth Tracking and Data Relay Satellite (TDRS). A secondary mission was to carry into orbit the Hitchhiker Diffuse X-ray Spectrometer and conduct experiments with it in the Shuttle payload bay. The author describes crew members, daily activities, launch preparation, and payloads during the seven-day mission.

Schuiling, Roelof. "STS-55 Mission Report." *Spaceflight*, July 1993. This article discusses STS-55, Space Shuttle *Columbia*'s mission to carry into orbit the reusable German Spacelab D-2 and to use it in the Shuttle's payload bay to conduct 88 experiments in astronomy, atmospheric physics, Earth observations, materials and life sciences, and technology applications. The author describes crew members, daily activities, launch preparation, and payloads during the 11-day mission.

Schuiling, Roelof. "STS-56 Mission Report." *Spaceflight*, June 1993. This article discusses STS-56, Space Shuttle *Discovery*'s mission to carry Atmospheric Laboratory for Applications and Science-2 (ATLAS-2) into orbit and to use it to conduct experiments in the Shuttle payload bay. ATLAS-2 was designed to collect data on the relationship between the Sun's energy output and Earth's middle atmosphere. The author describes crew members, daily activities, launch preparation, and payloads during the 10-day mission.

Schuiling, Roelof. "STS-57 Mission Report." *Spaceflight*, September 1993. This article discusses STS-57, Space Shuttle *Endeavour*'s mission to carry Spacehab into orbit and to use it to conduct biomedical and materials science experiments in the Shuttle payload bay. On this mission, Shuttle astronauts also conducted a spacewalk to retrieve the European Retrievable Carrier and stow it in the Shuttle's payload bay for return to Earth. The author describes crew members, daily activities, launch preparation, and payloads during the 11-day mission.

Schuiling, Roelof. "STS-58 Mission Report." *Spaceflight*, January 1994. This article discusses STS-58, Space Shuttle *Columbia*'s mission to carry into orbit the second dedicated Spacelab for Life Sciences and to use it to conduct experiments in the Shuttle payload bay. Shuttle crew used Spacelab to conduct 14 experiments in cardiovascular and cardiopulmonary physiology, musculoskeletal physiology, neuroscience, and regulatory physiology. The author describes launch crew members, daily activities, launch preparation, and payloads during the 15-day mission.

Schuiling, Roelof. "STS-59 Mission Report." *Spaceflight*, August 1994. This article discusses STS-59, Space Shuttle *Endeavour*'s mission to carry the Space Radar Laboratory (SRL) into orbit and to use it to conduct experiments in the Shuttle payload bay. The SRL, containing instruments such as the Spaceborne Imaging Radar-C, was designed to study Earth's ecosystem. The author describes crew members, daily activities, launch preparation, and payloads during the 12-day mission.

Schuiling, Roelof. "STS-60 Mission Report." *Spaceflight*, April 1994. This article describes STS-60, the first mission of the U.S.-Russian Shuttle-*Mir* program and the first Space Shuttle flight of a Russian cosmonaut, Sergei K. Krikalev. STS-60 carried into orbit the commercially developed Spacehab laboratory module and used it to conduct experiments in the Shuttle payload bay. The author describes crew members, daily activities, launch preparation, and payloads during the nine-day mission.

Schuiling, Roelof. "STS-62 Mission Report." *Spaceflight*, June 1994. This article discusses STS-62, Space Shuttle *Columbia*'s mission to carry into orbit NASA's U.S. Microgravity Payload 2 (USMP-2) and Office of Aeronautics and Space Technology 2 (OAST-2) and to use them to conduct experiments in the Shuttle payload bay. The author describes crew members, daily activities, launch preparation, and payloads during the 15-day mission.

Schuiling, Roelof. "STS-63 Mission Report." *Spaceflight*, May 1995. This article discusses STS-63, the second mission of the U.S.-Russian Shuttle-*Mir* program. In STS-63, *Discovery* carried out the first rendezvous of a U.S. Space Shuttle with Russia's space station *Mir*. Known as the Near-*Mir* mission, STS-63 carried into orbit Spacehab 3 and the Space Radar Laboratory (SRL). STS-63 was the first flight in which a woman—astronaut Eileen M. Collins—piloted the Shuttle, and the second Shuttle flight carrying a Russian cosmonaut—Vladimir G. Titov. The author describes the nine-day mission's crew members, daily activities, launch preparation, and payloads.

Schuiling, Roelof. "STS-64 Mission Report." *Spaceflight*, December 1994. The author of this article was Payload Manager of STS-64. In this article covering that mission, the author describes the preparation of Space Shuttle *Discovery* for launch and control room activities on launch day, focusing on the testing and preparation of STS-64's payloads, including the Lidar In Space Technology Experiment (LITE).

Schuiling, Roelof. "STS-64 Mission Report: Laser Atmospheric Research, Robotic Operations, Untethered Spacewalk." *Spaceflight*, December 1994. This article discusses STS-64, Space Shuttle *Discovery*'s mission to conduct atmospheric research with a laser, use a robot to process semiconductor materials, and perform an untethered spacewalk. The author describes preflight processing, crew members, and Shuttle activities for each of the 12 days that the Shuttle was in space.

Schuiling, Roelof. "STS-65 Mission Report." *Spaceflight*, October 1994. This article discusses STS-65, Space Shuttle *Columbia*'s mission to carry into orbit the second flight of the International Microgravity Laboratory 2 (IML-2). The author describes crew members, daily activities, launch preparation, and payloads during the 16-day mission.

Schuiling, Roelof. "STS-66 Mission Report." *Spaceflight*, March 1995. This article discusses STS-66, the mission of Space Shuttle *Atlantis* to carry into orbit seven instruments on the Atmospheric Laboratory for Applications, as well as the Science-3 Cryogenic Infrared Spectrometers and Telescopes for the Atmosphere-Shuttle Pallet. The author describes crew members, daily activities, launch preparation, and payloads during the 12-day mission.

Schuiling, Roelof. "STS-67 Mission Report." *Spaceflight*, July 1995. This article discusses STS-67, Space Shuttle *Endeavour*'s mission to carry into orbit the ASTRO Observatory and its three ultraviolet telescopes: the Hopkins Ultraviolet Telescope, the Wisconsin Ultraviolet Photo-Polarimeter Experiment, and the Ultraviolet Imaging. The author describes crew members, daily activities, launch preparation, and payloads during the 17-day mission.

Schuiling, Roelof. "STS-68 Mission Report." *Spaceflight*, February 1995. This article discusses STS-68, Space Shuttle *Endeavour*'s mission to carry into orbit, for the second time, the Space Radar Laboratory (SRL) and to use it to conduct experiments in the Shuttle payload bay. The author describes STS-68 crew members, daily activities, launch preparation, and payloads during the 12-day mission.

Schuiling, Roelof. "STS-69 Mission Report." *Spaceflight*, December 1995. This article discusses STS-69, Space Shuttle *Endeavour*'s mission to deploy and retrieve Spartan 201-03 and the Wake Shield Facility 2 (WSF-2), an experimental science platform. The author describes crew members, daily activities, launch preparation, and payloads during the 12-day mission.

Schuiling, Roelof. "STS-70 Mission Report." *Spaceflight*, November 1995. This article discusses STS-70, Space Shuttle *Discovery*'s mission to deploy the Tracking and Data Relay

Satellite-G (TDRS-G) and to conduct experiments such as the Biological Research in Canister, which investigates the effects of spaceflight on small arthropod animal and plant specimens. The author describes crew members, daily activities, launch preparation, and payloads during the 10-day mission.

Schuiling, Roelof. "STS-71 Mission Report." *Spaceflight*, October 1995. This article discusses STS-71, the mission of Space Shuttle *Atlantis* to dock with Russia's space station *Mir* and to carry out joint on-orbit operations. This mission was the first time that a U.S. Space Shuttle had docked at *Mir*. The author describes crew members, daily activities, launch preparation, and payloads during the 11-day mission.

Schuiling, Roelof. "STS-72 Mission Report." *Spaceflight*, April 1996. This article discusses STS-72, Space Shuttle *Endeavour*'s mission to retrieve the Japanese Space Flyer Unit satellite and to deploy and retrieve the Office of Aeronautics and Space Technology Flyer spacecraft. The author describes crew members, daily activities, launch preparation, and payloads during the 10-day mission.

Schuiling, Roelof. "STS-73 Mission Report." *Spaceflight*, January 1996. This article discusses STS-73, Space Shuttle *Columbia*'s mission to carry the second United States Microgravity Laboratory (USML-2) into space for microgravity studies in the Shuttle payload. The author describes crew members, daily activities, launch preparation, and payloads during the 17-day mission.

Schuiling, Roelof. "STS-74 Mission Report." *Spaceflight*, March 1996. This article discusses STS-74, the mission of Space Shuttle *Atlantis* to dock with Russia's space station *Mir* for the second time and to transfer equipment and supplies to *Mir*. The author describes crew members, daily activities, launch preparation, and payloads during the nine-day mission.

Schuiling, Roelof. "STS-75 Mission Report." *Spaceflight*, June 1996. This article discusses STS-75, Space Shuttle *Columbia*'s mission to deploy and retrieve the joint U.S.-Italian Tethered Satellite System. The crew had deployed the satellite and had begun gathering scientific data when the tether snapped on flight day three, as the satellite was just short of full deployment by about 12.8 miles. The author describes crew members, daily activities, launch preparation, and payloads during the 15-day mission.

Schuiling, Roelof. "STS-76 Mission Report." *Spaceflight*, July 1996. This article discusses STS-76, the mission of Space Shuttle *Atlantis* to dock with Russia's space station *Mir* for the third time and to deliver the first American female astronaut—Shannon W. Lucid—to live on *Mir*. The author describes crew members, daily activities, launch preparation, and payloads during the 10-day mission.

Schuiling, Roelof. "STS-77 Mission Report." *Spaceflight*, August 1996. This article discusses STS-77, Space Shuttle *Endeavour*'s mission to carry into orbit the pressurized research module, Spacehab 4, and to use it to conduct experiments in agriculture, biotechnology, electronic materials, and polymers in the Shuttle payload bay. STS-77 crew also deployed and retrieved the Spartan-207 free flyer. The author describes crew members, daily activities, launch preparation, and payloads during the 11-day mission.

Schuiling, Roelof. "STS-78 Mission Report." *Spaceflight*, October 1996. This article discusses STS-78, Space Shuttle *Columbia*'s mission to carry into space the Life and Microgravity Spacelab, to use it to conduct experiments in the Shuttle payload bay, and to study the effects of long-duration spaceflight on human physiology. The author describes crew members, daily activities, launch preparation, and payloads during the 18-day mission.

Schuiling, Roelof. "STS-79 Mission Report." *Spaceflight*, February 1997. This article discusses STS-79, the mission of Space Shuttle *Atlantis* to dock with Russia's space station *Mir* for the fourth time. The Shuttle brought the Spacehab Double Module to support Shuttle-*Mir* activities. The author describes crew members, daily activities, launch preparation, and payloads during the 11-day mission.

Schuiling, Roelof. "STS-80 Mission Report." *Spaceflight*, March 1997. This article discusses STS-80, Space Shuttle *Columbia*'s mission to deploy and retrieve the Orbiting and Retrievable Far and Extreme Ultraviolet Spectrometer-Shuttle Pallet Satellite 2 and the Wake Shield Facility 3 (WSF-3), an experimental science platform. The author describes crew members, daily activities, launch preparation, and payloads during the 18-day mission.

Schuiling, Roelof. "STS-81 Mission Report." *Spaceflight*, May 1997. This article discusses STS-81, the mission of Space Shuttle *Atlantis* to dock with Russia's space station *Mir* for the fifth time. The author describes crew members, daily activities, launch preparation, and payloads during the 18-day mission.

Schuiling, Roelof. "STS-82 Mission Report." *Spaceflight*, June 1997. This article discusses STS-82, Space Shuttle *Discovery*'s mission to service the HST. The author describes crew members, daily activities, launch preparation, and payloads during the nine-day mission.

Schuiling, Roelof. "STS-83 Mission Report." *Spaceflight*, October 1997. This article discusses STS-83, Space Shuttle *Columbia*'s mission to carry into orbit the Microgravity Science Laboratory 1 (MSL-1) and to use it to conduct 19 materials science investigations in the Shuttle payload bay. The mission was cut short because of concern over erratic readings from some Shuttle fuel cells. The author describes crew members, daily activities, launch preparation, and payloads during the four-day mission.

Schuiling, Roelof. "STS-84 Mission Report." *Spaceflight*, September 1997. This article discusses STS-84, the mission of Space Shuttle *Atlantis* to dock with Russia's space station *Mir* for the sixth time. The author describes crew members, daily activities, launch preparation, and payloads during the nine-day mission.

Schuiling, Roelof. "STS-85 Mission Report." *Spaceflight*, November 1997. This article discusses STS-85, Space Shuttle *Discovery*'s mission to deploy and retrieve the Cryogenic Infrared Spectrometers and Telescopes for the Atmosphere-Shuttle Pallet Satellite-2, as well as to carry into space a number of payloads involving secondary experiments. The author describes crew members, daily activities, launch preparation, and payloads during the 11-day mission.

Schuiling, Roelof. "STS-86 Mission Report." *Spaceflight*, January 1998. This article discusses STS-86, the mission of Space Shuttle *Atlantis* to dock with Russia's space station *Mir* for the seventh time. The author describes crew members, daily activities, launch preparation, and payloads during the 10-day mission.

Schuiling, Roelof. "STS-87 Mission Report." *Spaceflight*, April 1998. This article discusses STS-87, Space Shuttle *Columbia*'s mission to conduct science experiments and to deploy and retrieve the SPARTAN-201-04 free-flyer, a Solar Physics Spacecraft. Experiments conducted in the Shuttle payload bay on the U.S. Microgravity Payload focused on combustion science, fundamental physics, and materials science. The author describes crew members, daily activities, launch preparation, and payloads during the 15-day mission.

Schuiling, Roelof. "STS-88: 'Unity' Module Delivered to Space Station." *Spaceflight*, March 1999. This article discusses STS-88, Space Shuttle *Endeavour*'s mission to begin construction of the ISS. The author describes crew members, daily activities, launch preparation, and payloads during the 13-day mission.

Schuiling, Roelof. "STS-89 Mission Report." *Spaceflight*, June 1998. This article discusses STS-89, Space Shuttle *Endeavour*'s mission to dock with the Russian space station *Mir* to deliver scientific equipment, logistical hardware, and water. The author describes crew members, daily activities, launch preparation, and payloads during the eight-day mission.

Schuiling, Roelof. "STS-90 Mission Report." *Spaceflight*, August 1998. This article discusses STS-90, Space Shuttle *Columbia*'s mission to study neuroscience. The *Columbia* payload included Neurolab and its 26 experiments related to the nervous system. The author describes crew members, daily activities, launch preparation, and payloads during the 15-day mission.

Schuiling, Roelof. "STS-91 Mission Report." *Spaceflight*, November 1998. This article discusses STS-91, Space Shuttle *Discovery*'s mission to dock with the Russian space station *Mir* and to deliver cargo, science experiments, and supplies. In addition, the crew moved long-term U.S. experiments that had been aboard *Mir* into *Discovery*'s mid-deck locker area. The author describes crew members, daily activities, launch preparation, and payloads during the nine-day mission.

Schuiling, Roelof. "STS-92: Discovery Completes 100th Shuttle Flight." *Spaceflight*, January 2001. This article describes STS-92, Space Shuttle *Discovery*'s mission to continue the construction of the ISS. The article describes launch preparation, payloads, crew members, and daily activities during the 14-day mission.

Schuiling, Roelof. "STS-93: Launch Delays: Problems on Ascent." *Spaceflight*, December 1999. This article discusses STS-93, Space Shuttle *Columbia*'s mission to deploy the Chandra X-ray Observatory. STS-93 was the first mission in Space Shuttle history to be commanded by a woman—Eileen M. Collins. The author describes crew members, daily activities, launch preparation, and payloads during the six-day mission.

Schuiling, Roelof. "STS-94 Mission Report." *Spaceflight*, October 1997. This article discusses STS-94, Space Shuttle *Columbia*'s repeat mission to carry into orbit the Microgravity Science Laboratory 1 (MSL-1) after erratic fuel cell readings shortened MSL-1's trip aboard STS-83. The author describes crew members, daily activities, launch preparation, and payloads during the 15-day mission.

Schuiling, Roelof. "STS-95: 'The John Glenn Flight'." *Spaceflight*, February 1999. This article discusses STS-95, Space Shuttle *Discovery*'s mission to carry into orbit the pressurized Spacehab module and to use it to conduct a variety of scientific experiments in the Shuttle payload bay. On that mission, Shuttle crew members also deployed and retrieved the Spartan free-flyer payload, which gathered measurements of the solar corona and solar wind. Former Mercury program astronaut and U.S. Senator John H. Glenn Jr. was one of the STS-95 crew members, returning to space at the age of 77. The author describes the 10-day mission's crew members, daily activities, launch preparation, and payloads.

Schuiling, Roelof. "STS-96: Discovery Shuttle Launched on First Mission of 1999." *Spaceflight*, October 1999. This article discusses STS-96, Space Shuttle *Discovery*'s mission to deliver parts and materials to the ISS. The author describes crew members, day-to-day operations, launch preparation, and payloads during the nine-day mission.

Schuiling, Roelof. "STS-97: Endeavour Delivers Solar Arrays to ISS." *Spaceflight*, March 2001. This article discusses STS-97, Space Shuttle *Endeavour*'s mission to deliver solar arrays and supplies to the ISS. The author describes crew members, daily activities, launch preparation, and payloads during the 12-day mission.

Schuiling, Roelof. "STS-98: Atlantis Carries Destiny to Orbit." *Spaceflight*, May 2001. This article describes STS-98, the mission of Space Shuttle *Atlantis* to take the United States' laboratory Destiny to the ISS. The article names the crew and describes their activities for each of the 14 days the Shuttle was in space.

Schuiling, Roelof. "STS-99: Mapping the Earth's Surface by Radar." *Spaceflight*, May 2000. This article discusses STS-99, Space Shuttle *Endeavour*'s mission to map the Earth's surface using radar. The author describes the mission payload, which included the Shuttle Radar Topography Mission, the STS-99 crew, and the crew's activities on each of the 12 days that *Endeavour* was in space.

Schuiling, Roelof. "STS-100: Endeavour Carries Robotic Components to ISS." *Spaceflight*, August 2001. This article discusses STS-100, Space Shuttle *Endeavour*'s mission to equip and supply the ISS. The author describes crew members, daily activities, launch preparation, and payloads during the 13-day mission.

Schuiling, Roelof. "STS-101: Maintenance Mission to Unity-Zarya." *Spaceflight*, August 2000. This article discusses STS-101, the mission of Space Shuttle *Atlantis* to maintain the Unity and Zarya. Unity and Zarya were the first ISS components. The author describes crew members, daily activities, launch preparation, and payloads during the 11-day mission.

Schuiling, Roelof. "STS-102: Discovery Delivers Second Crew to Space Station." *Spaceflight*, June 2001. This article discusses STS-102, Space Shuttle *Discovery*'s mission to equip and supply the ISS. The author describes crew members, daily activities, launch preparation, and payloads during the 14-day mission.

Schuiling, Roelof. "STS-104: Atlantis Delivers Space Doorway to ISS." *Spaceflight*, October 2001. This article describes STS-104, the mission of Space Shuttle *Atlantis* to deliver and install a space doorway on the ISS. The article describes launch preparation, payloads, crew members, and activities during the 14-day mission.

Schuiling, Roelof. "STS-105: Space Shuttle Mission Report." *Spaceflight*, November 2001. This article discusses STS-105, Space Shuttle *Discovery*'s mission to equip and supply the ISS. The author describes crew members, daily activities, launch preparation, and payloads during the 13-day mission.

Schuiling, Roelof. "STS-106: Atlantis Revisits Space Station." *Spaceflight*, December 2000. This article describes STS-106, the mission of Space Shuttle *Atlantis* to equip the ISS and prepare the station for its first crew. The article describes launch preparation, payloads, crew members, and activities during the 13-day mission.

Schuiling, Roelof. "STS-107: Columbia's Final Mission." *Spaceflight*, April 2003. This article discusses Space Shuttle *Columbia*'s final mission, STS-107, describing launch preparation, payloads, crew members, and daily activities during the 15-day mission to conduct scientific experiments. *Columbia* and its crew were lost during reentry over east Texas on 1 February 2003, at about 9:00 a.m. (EST).

Schuiling, Roelof. "STS-108: Endeavour Carries Fourth Crew to Space Station." *Spaceflight*, March 2002. This article discusses STS-108, Space Shuttle *Endeavour*'s mission to equip and supply the ISS. The author describes crew members, daily activities, launch preparation, and payloads during the 13-day mission.

Schuiling, Roelof. "STS-111: Weather Delays Endeavour's Launch and Landing." *Spaceflight*, September 2002. This article discusses STS-111, Space Shuttle *Endeavour*'s mission to equip and supply the ISS. The author describes crew members, daily activities, launch preparation, and payloads during the 15-day mission.

Schuiling, Roelof. "STS-112: Advances Space Station Assembly." *Spaceflight*, January 2003. This article describes STS-112, the mission of Space Shuttle *Atlantis* to deliver to the ISS the S1 integrated truss segment and spacewalk platform and to install the new components. The article describes launch preparation, payloads, crew members, and activities during the 12-day mission.

Schuiling, Roelof. "STS-113: Endeavour Overcomes Technical Hitches To Deliver Sixth ISS Crew." *Spaceflight*, March 2003. This article discusses STS-113, Space Shuttle *Endeavor*'s mission to equip and supply the ISS. The author describes crew members, daily activities, launch preparation, and payloads during the 15-day mission.

Shayler, David J. *Walking in Space*. New York: Springer-Praxis, 2004. This book provides a comprehensive overview and analysis of the techniques astronauts use in EVAs (spacewalks). The author draws on original documentation and on personal interviews with astronauts who have EVA experience, as well as on the accounts of staff involved in spacesuit design, EVA planning, and operations. The book describes the development of techniques for ensuring crew safety during spacewalks and looks ahead to future EVAs from the ISS and the development of new technology.

Simpson, Clive. "STS-114: Return of the Space Shuttle." *Spaceflight*, October 2005. This article discusses STS-114, Space Shuttle *Discovery*'s mission to deliver equipment and supplies to the ISS. The author describes crew members, daily activities, launch preparation, and payloads during the 13-day mission.

Simpson, Clive. "STS-115: Astronauts Complete Tough Mission." *Spaceflight*, November 2006. This article discusses STS-115, the mission of Space Shuttle *Atlantis* to deliver to the ISS the P3/P4 integrated truss and to install the truss and a pair of solar arrays on the ISS. The author describes crew members, daily activities, launch preparation, and payloads during the 11-day mission.

Simpson, Clive. "STS-117: Atlantis Completes Spectacular Mission." *Spaceflight*, August 2007. This article discusses STS-117, the mission of Space Shuttle *Atlantis* to deliver to the ISS and to install the second and third starboard truss segments and an additional pair of solar arrays. The author describes crew members, daily activities, launch preparation, and payloads during the 17-day mission.

Simpson, Clive, Gerard van der Haar, and Rudolf van Beest. "STS-116: Shuttle Mission Re-Wires ISS." *Spaceflight*, February 2007. This article discusses STS-116, Space Shuttle *Discovery*'s mission to deliver equipment and supplies to the ISS. The author describes crew members, daily activities, launch preparation, and payloads during the 12-day mission.

Simpson, Clive, Gerard van der Haar, and Rudolf van Beest. "STS-118: Endeavour Finally Returns to Space." *Spaceflight*, October 2007. This article discusses STS-118, Space Shuttle *Endeavour*'s mission to deliver equipment and supplies to the ISS. The crew delivered and installed a third starboard truss segment and 5,000 pounds (2,268 kilograms) of equipment and supplies. The author describes crew members, daily activities, launch preparation, and payloads during the 12-day mission.

"The U.S. Space Shuttle: A Day in the Life." *Ad Astra*, August 2002. This article presents photographs of the life cycle of a Space Shuttle mission.

U.S. Congress. Senate. Committee on Commerce, Science, and Transportation. Subcommittee on Science, Technology, and Space. *Results of Space Shuttle Endeavour Mission.* 102nd Cong., 2nd sess., 11 June 1992. *https://web.lexis-nexis.com/congcomp/* (accessed 11 November 2011 via Proquest Congressional). This hearing reviews the results of the 1992 Space Shuttle *Endeavour* flight, STS-49, specifically on-board research activities and the retrieval and reboost of a communications satellite.

Van de Haar, Rudolf. "STS-126: Mission Doubles Up Crew Capacity." *Spaceflight*, February 2008. This article discusses STS-126, Space Shuttle *Endeavour*'s mission to deliver construction equipment to the ISS and to service the Solar Alpha Rotary Joints. The author describes crew members, daily activities, launch preparation, and payloads during the 15-day mission.

Van de Haar, Rudolf, Rudolf van Beest, and Clive Simpson. "STS-121: Back to Back for Discovery." *Spaceflight*, September 2006. This article discusses STS-121, Space Shuttle *Discovery*'s mission to deliver equipment and supplies to the ISS and to demonstrate techniques for inspecting and protecting the Shuttle's thermal protection system. The author describes crew members, daily activities, launch preparation, and payloads during the 12-day mission.

Wilson, Keith T. "From Landing . . . To Launch." *Spaceflight*, October 1997. This article describes the short turnaround time (94 days) between Space Shuttle *Columbia*'s STS-83 and STS-94 missions, as well as the missions' prelaunch processing. The staff of NASA's Kennedy Space Center staff tested and evaluated the Shuttle for its return to space. The Microgravity Science Laboratory remained in the Shuttle between the two flights, the first time that a major payload had remained on-board while the Shuttle was on the ground.

Link to Part 1 (1970–1991), Chapter 6—Space Shuttle Operations

CHAPTER 6—*CHALLENGER* ACCIDENT AND AFTERMATH

Adamson, Heather. *The Challenger Explosion*. Mankato, MN: Capstone Press, 2006. Ages 9–12. This graphic novel tells the story of Christa McAuliffe and the six other NASA astronauts who lost their lives in the Space Shuttle *Challenger* disaster on 28 January 1986.

Arnett, Glenn. "Search for the *Challenger*." *Air & Space*, May 1994. The article reports on the author's participation in the search for Space Shuttle *Challenger*. The author describes difficulties in conducting the search, diving systems used in the search, the discovery of the main debris area, the discovery of *Challenger* crews' remains, and how the search affected the author.

Aviation Week Video. *Space Shuttle the Recovery*. Videocassette (VHS). New York: McGraw-Hill, 1992. This video focuses on the redesign and rebuilding of the Space Shuttle program at NASA to pinpoint exactly what went wrong with *Challenger* and how the problems were investigated and rectified.

Biggs, R. E. *Space Shuttle Main Engine: The First Twenty Years and Beyond*. American Astronautical Society History Series, vol. 29. San Diego, CA: Univelt, 2008. This article describes the Space Shuttle's main engine design and development, outlining requirements, obstacles, component testing, and engine testing. Discussing the *Challenger* tragedy and the Space Shuttle's return to flight, the author recounts NASA's efforts to attain full-power-level certification and acceptance tests for the Shuttle.

Bredeson, Carmen. *The Challenger Disaster: Tragic Space Flight*. Berkeley Heights, NJ: Enslow Publishers, 1999. Ages 9–12. This book describes the events surrounding the explosion of Space Shuttle *Challenger* in 1986 and the investigation of the disaster, as well as the stories of the seven astronauts who died on the *Challenger*.

Burgess, Colin. "Lost Mission." *Spaceflight*, June 1999. This article describes how the loss of Space Shuttle *Challenger* in 1986 affected the future missions of Indonesian astronaut candidates for Space Shuttle missions. The U.S. government offered to train an Indonesian astronaut who would be a crew member on a mission to launch an Indonesian satellite. Post-tragedy launch delays forced NASA to launch the Indonesian satellite on an unmanned booster so the Indonesian astronaut candidates did not get to fly a Shuttle mission. The article also portrays the two astronaut candidates chosen to for the mission, Taufik Akbar and Pratiwi Sudarmono.

Caper, William. *The Challenger Space Shuttle Explosion*. New York: Bearport, 2007. This book examines the events surrounding the explosion of Space Shuttle *Challenger* and discusses the significant changes to the Space Shuttle program that resulted from the loss of *Challenger*.

Chiles, James R. "Out from the Shadow." *Air & Space*, May 1996. This article profiles the Thiokol Corporation, the maker of the solid rockets that propel the Space Shuttle into orbit. The author discusses the loss of Space Shuttle *Challenger* in 1986 and the role of the solid rocket in the *Challenger*'s explosion.

Cole, Michael D. *Challenger: America's Space Tragedy*. Springfield, NJ: Enslow Publishers, 1995. Ages 9–12. Grades 4–6. The author clearly describes the Space Shuttle's technical problems, turning each mission into a tense tale of danger, courage, and obstacles overcome or—in *Challenger*'s case—deliberately ignored. The book, which includes the biographies of the people involved with *Challenger*, includes many small photos.

Evans, Ben. *Space Shuttle Challenger: Ten Journeys into the Unknown*. Chichester, UK: Praxis Publishing, 2007. This book details the design, development, and construction of Space Shuttle *Challenger*. There are stories of *Challenger*'s missions from the points of view of the astronauts, engineers, scientists who flew and knew the Shuttle, and the managers, technicians, and ground personnel who worked on *Challenger* and made the spacecraft into one of the most capable Shuttles in NASA's service. *Challenger* veterans, including C. Gordon Fullerton and Vance D. Brand, describe their experiences and the differences between *Challenger* and the other Shuttles.

Fahey, Kathleen. *Challenger and Columbia*. Milwaukee, WI: Gareth Stevens Publishing, 2005. Ages 9 and older. The book describes the events leading up to the loss of Space Shuttles *Challenger* and *Columbia*. Based on detailed descriptions, expert testimony, and firsthand accounts, the author takes an in-depth look at the *Challenger* and *Columbia* tragedies and discusses the lessons learned from them.

Hall, Joseph Lorenzo. "Columbia and Challenger: Organizational Failure at NASA." *Space Policy* 19, no. 4 (November 2003): 237–247. This article outlines some of the critical features of NASA's organization. The author discusses organizational change at NASA, namely "path dependence" and "normalization of deviance." He reviews the reasons that experts have called the *Challenger* tragedy an organizational failure. Finally, he argues that the recent *Columbia* accident also displays characteristics of organizational failure and proposes recommendations for the future.

Heimann, C. F. Larry. "Understanding the Challenger Disaster: Organizational Structure and the Design of Reliable Systems." *American Political Science Review* 87, no. 2 (June 1993): 421–435. This article describes the loss of Space Shuttle *Challenger* and assesses the effect that the disaster had on NASA's organizational ability and behavior. The author presents a theory of organizational reliability before and after the incident.

Holden, Henry M. *The Tragedy of the Space Shuttle Challenger*. Berkeley Heights, NJ: MyReportLinks.com Books, 2004. Ages 9–12. This book describes events surrounding the explosion of Space Shuttle *Challenger* in 1986 and discusses the investigation of this disaster. The author tells the stories of the seven astronauts who died. The book also includes Internet links to related Web sites, source documents, and photographs.

Kortenkamp, Steve. *Space Shuttles*. Mankato, MN: Capstone Press, 2008. Ages 4–8. This book describes the history and uses of NASA's Space Shuttles, including both the practical uses and the drawbacks of reusable Space Shuttles. The author explains how the Shuttle flies and how it launches satellites. In addition, he describes how astronauts travel into space to make repairs on missions such as that to service the HST. The book also includes information about the two Shuttle disasters.

Lieurance, Suzanne. *The Space Shuttle Challenger Disaster in American History*. Berkeley Heights, NJ: Enslow Publishers, 2001. Ages 9–12. Grades 5–7. After recounting the events of the morning that Space Shuttle *Challenger* exploded, the author places the *Challenger*'s mission in historical context. She summarizes the U.S. space program up to that time, profiles each of the astronauts assigned to the *Challenger* crew, and discusses the selection and training of civilian teacher Christa McAuliffe for the mission.

Maier, Mark. "Ten Years After a Major Malfunction . . . : Reflections on 'The *Challenger* Syndrome'." *Journal of Management Inquiry* 11, no. 3 (September 2002): 282–292. This article reviews the insights gleaned from the *Challenger* disaster case study. Little known facts about the case are revealed, including serious questions about the veracity of the Rogers Commission report, which investigated the accident. Seven key elements of "the *Challenger* syndrome" are presented, pinpointing existing bureaucratic imperatives as a primary obstacle to the exercise of leadership and ethical decision making. The article closes with 10 central leadership lessons from the *Challenger* case study.

Martin, Ryan M, and Louis A. Boynton. "From Liftoff to Landing: NASA's Crisis Communications and Resulting Media Coverage Following the Challenger and Columbia Tragedies." *Public Relations Review* 31, no. 2 (June 2005): 253–261. NASA's public relations effort following the explosion of *Challenger* in 1986 is considered an example of crisis communications failure. However, after the *Columbia* disaster in 2003, NASA was praised for its successful handling of the crisis. Using widely accepted crisis communication concepts associated with stakeholder theory, the author discusses how four newspapers presented NASA's crisis communication efforts following the two crises, showing that the print media accorded NASA more positive coverage following the *Columbia* disaster than the *Challenger* disaster.

McDonald, Allan J., and James R. Hansen. *Truth, Lies, and O-Rings: Inside the Space Shuttle Challenger Disaster*. Gainesville, FL: University Press of Florida, 2009. The authors are an engineer and executive of Morton Thiokol, the maker of the Shuttle's solid rocket booster, and aerospace historian James R. Hansen. The book draws attention to the factors that led to the accident, some of which were never included in NASA's Failure Team report submitted to the Presidential Commission. It also addresses the early warnings of very severe debris issues from the first two post-*Challenger* flights.

McNeese, Tim. *The Challenger Disaster*. New York: Children's Press, 2003. This book documenting the *Challenger* disaster chronicles the failures on the part of NASA engineers to prevent the tragic accident and provides information about the mission and its crew members.

Morring, Frank, Jr. "Risk Analyses; Academies Panel Urges NASA To Use Shuttle Instead of Robot for Hubble Servicing." *Aviation Week & Space Technology*, 13 December 2004. The author of this article urges NASA to send a Space Shuttle mission to service the HST, arguing that it is too valuable to risk sending a robotic mission to extend the Hubble's life. In the wake of the *Challenger* and *Columbia* disasters, NASA Administrator Sean O' Keefe had cancelled the planned Shuttle mission to the Hubble out of concern for the potential risk to the Shuttle crew, but a panel of experts finds that recent safety upgrades to the Shuttle have made the risk of a mission to the HST only marginally greater than the planned missions to the ISS.

Pike, John. "But What Is the True Rationale of Human Spaceflight?" *Space Policy* 10, no. 3 (August 1994): 217–222. The author states that the U.S. space program was adrift after the *Challenger* accident. He then focuses on the need for human spaceflight and its contribution to Space Shuttle projects that would benefit society.

Powell, Joel W. "A Cold Day at the Kennedy Space Center." *Spaceflight*, February 2006. This article marks the twentieth anniversary of the *Challenger* disaster, which occurred on 28 January 1986. The author describes liftoff preparations for the days leading up to the launch. It would take 32 months before NASA would resume Shuttle flights after the *Challenger* disaster.

Shayler, David. *Disasters and Accidents in Manned Spaceflight*. London: Springer, 2000. This book provides a superb overview of the history of human spaceflight accidents to the date of its publication. After explaining the challenges and dangers of launch, flight, and landing, the author discusses and evaluates each major spaceflight accident, as well as lesser known near accidents. The loss of *Columbia* had not yet occurred when the book was published.

Spangenburg, Ray, Diane Moser, and Kit Moser. *Onboard the Space Shuttle*. New York: Franklin Watts, 2002. Grades 5–9. The authors explain what life in space is like and describe the oddities that an aspiring astronaut must expect. Sections of the book cover the significant accomplishments of landmark Space Shuttle missions, such as the mission to repair the HST, as well as space stations and the *Challenger* explosion. The book also features many interesting color photographs and sidebars providing statistics, information about individual astronauts, and scientific principles related to space travel.

Streissguth, Thomas. *The Challenger: The Explosion on Liftoff*. Mankato, MN: Capstone High-Interest Books, 2003. Ages 9–12. This book examines Space Shuttle *Challenger* and the events that led to its destruction, along with the effects of the disaster on NASA's space program.

Tompkins, Phillip K. *Apollo, Challenger, Columbia: The Decline of the Space Program: A Study in Organizational Communication*. New York: Oxford University Press, 2004. This book portrays NASA from 1958 to 2003, concentrating on several specific points in the history of the space program, including 1986 and 2003, the periods surrounding the two Space Shuttle disasters. The author investigates the internal communication failures that resulted in the 1986 explosion of *Challenger* and the catastrophic failure of *Columbia* in 2003.

U.S. General Accounting Office. "Space Shuttle: Need To Sustain Launch Risk Assessment Process Improvements." Report no. GAO/NSIAD-96-73, Washington, DC, March 1996. *http://www.gao.gov/archive/1996/ns96073.pdf* (accessed 8 March 2012). This report reviews NASA's actions to improve the flow of information in launch decisions following the *Challenger* explosion. The GAO found that NASA has successfully created numerous formal and informal communication channels and an open organizational culture that encourages people to discuss safety concerns and to elevate unaddressed concerns to higher management levels.

Vaughan, Diane. *The Challenger Launch Decision: Risky Technology, Culture, and Deviance at NASA*. Chicago: University of Chicago Press, 1996. The author recreates the steps leading up to the fateful decision to launch *Challenger*, contradicting conventional interpretations to prove that what occurred at NASA was not misconduct but a disastrous mistake. She reveals how and why NASA insiders, when repeatedly faced with evidence that something was wrong, normalized the deviance so that it became acceptable to them.

White, Thomas Gordon, Jr. "The Establishment of Blame as a Framework for Sensemaking in the Space Policy Subsystem: A Study of the Apollo 1 and Challenger Accidents—Open Thesis." PhD Dissertation, Virginia Polytechnic Institute and State University, Blacksburg, VA, April 2000. *http://www.openthesis.org/documents/Establishment-Blame-As-Framework-Sensemaking-582553.html* (accessed 8 March 2012). This dissertation investigates how the establishment of blame becomes a framework for a national policy subsystem to make sense of a tragic event. Using the Apollo 1 and Space Shuttle *Challenger* accidents as case studies, the author examines the space policy subsystem's response to these two accidents and the process of establishing culpability.

Link to Part 1 (1970–1991), Chapter 7—Challenger Accident and Aftermath

CHAPTER 7—THE SPACE SHUTTLE AND THE HUBBLE SPACE TELESCOPE

"A Black Hole in the Sky." *Economist*, 14 November 1998. NASA is upgrading the Space
Shuttle fleet to enable the Shuttles to carry parts needed to build the ISS, thereby
reducing the space station's dependence on unpiloted Russian supply ships for refueling.
NASA may delay essential repairs to the HST so that it can send additional Shuttle flights
to help build the ISS.

Achenbach, Joel. "Hubble Mission Opens Shuttle's Last Act; Aging, Flawed Space Vehicle Still
Has Its Fans." *Washington Post*, 12 May 2009. As the end of the Space Shuttle program
draws near, this article discusses its use in supporting the HST.

Covault, Craig. "Hubble Mission Scrambles To Make Surprise Repairs." *Aviation Week & Space
Technology*, 24 February 1997. This article describes EVAs (spacewalks) that the Space
Shuttle crew performed during STS-82 to repair the HST and explains how the crew
overcame major unexpected events that occurred during that mission.

Cowen, Ron. "End of the Line for Hubble?" *Science News*, July 24, 2004. Citing safety concerns
that came to light after the Space Shuttle *Columbia* tragedy in February 2003, NASA
Administrator Sean O'Keefe announced that NASA was not going to send any more
Shuttle missions to upgrade or repair the HST. Rather than sending astronauts, NASA
proposes to send a robotic mission to repair and upgrade the orbiting observatory.

DiChristina, Mariette. "Fixing Hubble." *Popular Science*, December 1993. This article focuses
on NASA's 11-day mission, STS-61, to repair the troubled HST.

"Double, Double, Hubble Trouble." *Economist*, 27 November 1999. This article discusses the
various problems that NASA encountered during the second half of 1999, including the
shutdown of the HST, as the result of gyroscope failures, and the grounding of NASA
Space Shuttles following problems with *Columbia*. The article also reports that NASA
has scheduled Space Shuttle *Discovery* for a 10-day mission to fix the HST.

Green, Andrew. "STS-125: Hubble's Final Refit in Orbit." *Spaceflight*, August 2009. This article
describes STS-125, Space Shuttle *Atlantis*'s mission to repair and upgrade the HST. The
article describes prelaunch activities, crew members, and daily activities during the 12-
day mission.

Gugliotta, Guy. "Use Shuttle To Fix Hubble, NASA Is Told; Risks to Astronauts Acceptable,
Panel Says." *Washington Post*, 9 December 2004. A congressionally mandated
committee of the National Research Council (NRC) delivered to NASA's leadership a
report concluding that NASA would be able to use the Space Shuttle to service the HST
without posing unacceptable risks to the astronauts.

"Hubble Repair Mission: STS-61." *Spaceflight*, January 1994. This short article describes STS-
61, Space Shuttle *Endeavour*'s successful mission to repair the HST. The article gives a
daily list of activities of the *Endeavour* crew.

Jenkins, Dennis R., and Jorge R. Frank. *Servicing the Hubble Space Telescope: Space Shuttle Atlantis—2009*. North Branch, MN: Specialty Press, 2009. This book describes STS-125, the Space Shuttle mission to service the HST for one last visit before the Shuttle fleet retires. Over 12 days and five spacewalks, the crew of Space Shuttle *Atlantis* made repairs and upgrades to the telescope, leaving it in better shape than ever and ready for another five years or more of research.

Macilwain, Colin. "Hubble Stakes Are High for NASA's Future." *Nature* 366, no. 6453 (25 November 1993): 290. This article describes a planned high-stakes Space Shuttle mission to repair the HST and discusses NASA's battle against budget cuts.

Morring, Frank, Jr. "A Power Play 315 Miles High." *Aviation Week & Space Technology*, 11 March 2002. In a difficult service mission, the crew from Space Shuttle *Columbia* upgraded instruments and installed two new solar arrays on the HST. The solar arrays will generate more power, while reducing atmospheric drag.

Morring, Frank, Jr. "One More Time; Hubble Program Planning Final Shuttle Mission as Early as 2007 To Maintain the Observatory." *Aviation Week & Space Technology*, 11 July 2005. This article describes NASA's decision to approve a Shuttle mission to repair the HST. The author also discusses NASA's efforts to schedule a realistic number of Shuttle flights to enable the ISS partners to complete assembly of the space station before the Shuttle's planned retirement at the end of 2010.

Morring, Frank, Jr. "Risk Analyses; Academies Panel Urges NASA To Use Shuttle Instead of Robot for Hubble Servicing." *Aviation Week & Space Technology*, 13 December 2004. The author of this article urges NASA to send a Space Shuttle mission to service the HST, arguing that it is too valuable to risk sending a robotic mission to extend the HST's life. In the wake of the *Challenger* and *Columbia* disasters, NASA Administrator Sean O'Keefe had cancelled the planned Shuttle mission to the HST out of concern for the potential risk to the Shuttle crew. However, a congressionally mandated committee of the National Research Council (NRC) has found that recent safety upgrades to the Shuttle have made the risk of a mission to the HST only marginally greater than the risks involved in the planned missions to the ISS.

Morring, Frank, Jr., and Jefferson Morris. "Paradigm Shift; Hubble Repair Mission Recast for a New Age of Exploration." *Aviation Week & Space Technology*, 6 November 2006. This article describes NASA's preparations for a Space Shuttle mission to service the HST.

Morgan, Daniel. "Hubble Space Telescope: NASA's Plans for a Servicing Mission." CRS Report for Congress, Congressional Research Service, Library of Congress, Washington, DC, 23 May 2008. *http://digital.library.unt.edu/ark:/67531/metacrs10603/* (accessed 8 March 2012). This report describes the background for NASA's decision to service the HST with a mission to replace key telescope components. Without the mission, the HST would cease scientific operations in 2008.

National Research Council. *Assessment of Options for Extending the Life of the Hubble Space Telescope; Final Report*. Washington, DC: The National Academies Press, 2005. *http://www.nap.edu/openbook.php?record_id=11169&page=R1* (accessed 10 April 2012). In this report, the Committee on the Assessment of Options for Extending the Life of the Hubble Space Telescope assesses the scientific value of continued HST operation, the safety issues of using the Space Shuttle and a crew of astronauts to service the HST, the feasibility of robotic servicing, the risks and benefits of acceptable servicing options, and the effects of these servicing options on the HST's science capability.

Pickrell, J. "Telescope Tuned Up." *Science News*, 16 March 2002. This article describes how the crew of Space Shuttle *Columbia* renovated the HST on 9 March 2002. Over five days, four STS-109 astronauts performed 36 hours of spacewalks from *Columbia* to replace worn components of the HST and to install new devices.

Schuiling, Roelof. "STS-61 Mission Report." *Spaceflight*, March 1994. This article describes STS-61, Space Shuttle *Columbia*'s first mission to service and add new solar arrays to the HST. The article describes launch preparation for the mission, payloads, crew members, and daily activities during the 12-day mission.

Schuiling, Roelof. "STS-82 Mission Report." *Spaceflight*, June 1997. This article discusses STS-82, Space Shuttle *Discovery*'s mission to service the HST. The author describes crew members, daily activities, launch preparation, and payloads during the nine-day mission.

Schuiling, Roelof. "STS-103: Three EVAs Fix Hubble." *Spaceflight*, March 2000. This article describes STS-103, Space Shuttle *Discovery*'s mission to service and repair the HST. The article describes launch preparation, payloads, crew members, and daily activities during the nine-day mission.

Schuiling, Roelof. "STS-109: Columbia Mission Upgrades Hubble." *Spaceflight*, June 2002. This article describes STS-109, Space Shuttle *Columbia*'s mission to service the HST. The article describes launch preparation, payloads, crew members, and daily activities during the 12-day mission.

Spangenburg, Ray, Diane Moser, and Kit Moser. *Onboard the Space Shuttle*. New York: Franklin Watts, 2002. Grades 5–9. The authors explain what life in space is like and describe the oddities that an aspiring astronaut must expect. Sections of the book cover the significant accomplishments of landmark Space Shuttle missions, such as the missions to repair the HST, as well as space stations and the *Challenger* explosion. The book also features many interesting color photographs and sidebars providing statistics, information about individual astronauts, and scientific principles related to space travel.

U.S. Congress. House of Representatives. Committee on Science and Technology. *NASA's Space Shuttle and International Space Station Programs: Status and Issues*. 110th Cong., 1st sess., 24 July 2007. *http://www.gpo.gov/fdsys/pkg/CHRG-110hhrg36737/html/CHRG-110hhrg36737.htm* (accessed 8 March 2012). This hearing examines the main challenges to NASA accomplishing its major goals—to continue successfully flying the Space

Shuttle until its planned retirement in 2010 and to complete a planned Shuttle mission to service the HST. The hearing also considers the obstacles to completing the assembly of the ISS by the time NASA retires the Space Shuttle.

U.S. Congress. Senate. Committee on Appropriations. Subcommittee on Veterans Affairs, Housing and Urban Development, and Independent Agencies. *Hubble Space Telescope: Special Hearing.* 103rd Cong., 2nd sess., 8 February 1994. *https://web.lexis-nexis.com/ congcomp/* (accessed 11 November 2011 via Proquest Congressional). This hearing reviews NASA's 1994 Space Shuttle mission STS-61 to repair the HST.

U.S. Government Accountability Office. "Space Shuttle: Costs for Hubble Servicing Mission and Implementation of Safety Recommendations Not Yet Definitive." Report no. GAO-05-34, Washington, DC, November 2004. *http://www.gao.gov/new.items/d0534.pdf* (accessed 8 March 2012). This report reviews NASA's decision to cancel the final planned HST servicing mission, a decision that prompted debate about potential alternatives to prolonging Hubble's mission. The GAO addresses the basis of NASA's cost estimates to service the HST using the Space Shuttle and to implement recommendations made by the *Columbia* Accident Investigation Board (CAIB).

Zubrin, Robert. "Ditching Hubble Bodes Poorly for Future Space Exploration." *Ad Astra*, June 2004. This article presents information on the decision of NASA Administrator Sean O'Keefe to halt all support missions to the HST on 16 January 2004. The article also notes HST's discoveries, the costs of the HST support missions, and descriptions of the risks involved with Shuttle flights.

Link to Part 1 (1970–1991), Chapter 9—Science on the Shuttle, Potential and Actual

CHAPTER 8—SCIENCE ON THE SPACE SHUTTLE

Burgess, Colin, and Chris Dubbs. *Animals in Space: From Research Rockets to the Space Shuttle*. New York: Springer, 2007. This book is a detailed account of the history of animal spaceflights carried out by the U.S.S.R., United States, and other nations. The book describes how animal high-altitude and spaceflight research affected spaceflight biomedicine and technology and helped human beings undertake spaceflight with greater understanding and confidence.

Covault, Craig. "Unusual Shuttle Operations Advance Commercialization." *Aviation Week & Space Technology*, 2 December 1996. This article describes STS-80 and the difficulty the *Columbia* crew had in releasing the 2-ton (1.8-tonne) commercial Wake Shield Facility (WSF), an experimental science platform, from the Shuttle payload. The WSF, used to produce wafer semiconductors in space, is an example of the Space Shuttle's commercial utility.

Davidsen, Arthur F. "Far-Ultraviolet Astronomy on the Astro-1 Space Shuttle Mission." *Science* 259, no. 5093 (15 January 1993): 327. This article examines early results from the Hopkins Ultraviolet Telescope, observations obtained on the Astro-1 mission, STS-35, launched on 2 December 1990 and landed on 10 December 1990, of Space Shuttle *Columbia*.

Evans, Ben. "STS-70 Preview." *Spaceflight*, May 1995. This article previews STS-70, Space Shuttle *Discovery*'s mission to deliver the Tracking and Data Relay Satellite (TDRS) and to conduct research with the Inter-Mars Tissue Equivalent Proportional Counter. The article describes the crew members and secondary payload experiments of STS-70.

Heusel, Catherine. "Medicine from Space; NASA's Efforts To Probe the Heavens Have Produced Down to Earth Medical Advances That Keep Fetuses Alive, Provide Early Diagnosis of Cancer, and Offer Hope for Treating Osteoporosis." *Washington Post*, 10 August 1999. Among more than a dozen scientific experiments, Space Shuttle *Columbia* crew carried out two medical tests to determine whether certain drugs might protect future space voyagers from the bone and muscle degeneration caused by weightlessness. These tests are part of NASA's long-term medical mission to monitor and treat people who are hundreds, thousands and, potentially, millions of miles away from the nearest hospital.

Kremer, Ken. "STS-132: Atlantis' Last Blast with Russian Beauty." *Spaceflight*, August 2010. This article describes STS-132, Space Shuttle *Atlantis*'s final mission before its retirement. The goal of STS-132 was to equip and supply the ISS. Besides carrying spare parts to the ISS, *Atlantis* delivered the 11,000-pound (4.989.5-kilogram) Russian Rassvet Mini Research Module, which will enable ISS crew to conduct biotechnology and fluid physics experiments. The article describes prelaunch activities, crew members, payloads, and daily operations during the 11-day mission.

Lawler, Andrew. "How Much Space for Science." *Science* 303, no. 5658 (30 December 2004): 610–612. The article describes a U.S. National Academy of Science conference for scientists and engineers who gathered to offer advice regarding the country's human spaceflight program. The article describes the struggle between those scientists and engineers who prefer human spaceflight, using the Space Shuttle and other space vehicles, and those who prefer unpiloted spaceflight.

Lawler, Andrew. "NASA May Cut Shuttle Flights and Reduce Science on Station." *Science* 309, no. 5734 (22 July 2005): 540–541. This article focuses on NASA's plan to curtail Space Shuttle flights and to reduce science research activities on the ISS.

Lawler, Andrew. "Will NASA Annihilate Station Antimatter Experiment?" *Science* 303, no. 5664 (12 March 2004): 1590–1591. This article indicates NASA is reconsidering its support for an innovative experiment that is designed to capture direct evidence of elusive antimatter. The Alpha Magnetic Spectrometer is designed to detect antimatter, which scientists believe makes up half of the universe but has never been found.

Lawler, Andrew, Daniel Clery, and Dennis Normile. "Life Science Research on Space Station Is Headed for Big Cuts." *Science* 308, no. 5722 (29 April 2005): 610–611. The authors report that NASA is finalizing a new plan to reduce the quality and quantity of cutting-edge research on the ISS. The cost of returning the Space Shuttle to flight, finishing the ISS by 2010, and building new launchers will have a negative effect on continuing high-priority research in biology at the space station.

Leary, Warren E. "Newts and Metal Projects Ride Columbia into Space." *New York Times*, 9 July 1994. Space Shuttle *Columbia* carried a crew of seven into space today for a two-week mission, STS-65, to study both the subtle and the grand effects of gravity on materials and living things. The experiments involve work with exotic furnaces that can produce unusual alloys and with a rotating centrifuge that can simulate different levels of gravity. The equipment also includes aquariums teeming with thousands of animals, including goldfish, killifish, jellyfish, sea urchins, and Japanese red-bellied newts.

Leary, Warren E. "Shuttle Soars into Space on Mission To Detect Environmental Changes on Earth." *New York Times*, 1 October 1994. During STS-68, Space Shuttle *Endeavour* began sweeping Earth with advanced radars in an effort to detect and monitor both natural and human-induced environmental changes that affect life on the planet. In addition to studying global change, the radars are to be used to examine the habitat of endangered pandas in China, to inspect the area around the Chernobyl nuclear power plant in Ukraine to see how the environment has recovered from the 1986 nuclear accident, and to scan stagnant water pools in areas of malaria outbreaks to see which might best support disease-carrying mosquito populations.

Leath, Kevin. "Shaping Up for the Final Frontier." *Ad Astra*, June 1993. This article outlines the life sciences research requirements and the needed ground-based or in-space platforms of supporting human space exploration. Crew health and life-friendly spacecraft environments are also described.

Lenorovitz, Jeffrey M. "Steady Growth Seen for Commercial Space." *Aviation Week & Space Technology*, 15 March 1993. In this article, the author argues that Space Shuttles will fly missions that include more science experiments. *Endeavour* has been fitted with Spacehab, a pressurized laboratory that provides additional locker volume, as well as working and living areas for astronauts and experiments. Spacehab will enable more materials processing and other production work in microgravity.

McElroy, John H. "Some Thoughts on Space Station Science." *Space Policy* 17, no. 4 (November 2001): 257–260. The author comments that ISS budget cuts negatively affect U.S. scientific capabilities and undermine the United States' investment in the Space Shuttle.

National Research Council. Committee on an Assessment of Balance in NASA's Science Programs. *An Assessment of Balance in NASA's Science Programs.* Washington DC: National Academies Press, 2006. In this report, the National Research Council (NRC) assesses the health of NASA's support for its established scientific disciplines, under the budget requests imposed by the space exploration initiative. The report also analyzes NASA's science budget to determine whether it reflects cross-disciplinary scientific priorities appropriately.

Pletser, Vladimir. "ESA's Fluid Physics Experiments on STS-78." *Spaceflight*, October 1996. This article describes the fluid physics experiments that ESA conducted aboard *Columbia* during STS-78 in 1996.

Schuiling, Roelof. "Loads of Payloads!" *Spaceflight*, August 1996. This article describes payloads aboard Space Shuttle *Endeavour* STS-77 in 1996. The article gives short descriptions of the many experiments on Spacehab 4 and other payload experiments such as the Spartan-207 Inflatable Antenna Experiment and the Technology Experiments for Advancing Missions in Space.

Schuiling, Roelof. "STS-45 Mission Report." *Spaceflight*, October 1992. This article describes STS-45, the mission of Space Shuttle *Atlantis* to carry the first Atmospheric Laboratory for Applications and Science (ATLAS 1) into orbit. ATLAS 1 was designed to conduct studies in atmospheric chemistry, solar radiation, space plasma physics, and ultraviolet astronomy. The article describes launch preparation, payloads, crew members, and activities during the nine-day mission.

Schuiling, Roelof. "STS-47 Mission Report." *Spaceflight*, April 1993. This article discusses STS-47, Space Shuttle *Endeavour*'s mission to carry Spacelab-J into orbit and to use it to conduct experiments in the Shuttle payload. A joint mission of NASA and Japan's space agency, NASDA, Spacelab-J uses an unpiloted Spacelab module to conduct microgravity investigations in materials and life sciences. The author describes launch preparation, payloads, crew members, and daily activities during the eight-day mission.

Schuiling, Roelof. "STS-53 Mission Report." *Spaceflight*, March 1993. This article discusses STS-53, Space Shuttle *Discovery*'s mission to deploy a classified DOD satellite and to conduct two unclassified secondary experiments in the Shuttle payload bay. The two secondary experiments were the Orbital Debris Radar Calibration Spheres and the Shuttle Glow Experiment/Cryogenic Heat Pipe Experiment. The author describes crew members, daily activities, launch preparation, and payloads during the eight-day mission.

Schuiling, Roelof. "STS-54 Mission Report." *Spaceflight*, March 1993. This article discusses STS-54, Space Shuttle *Endeavour*'s mission to deploy the fifth Tracking and Data Relay Satellite (TDRS). A secondary mission was to carry into orbit the Hitchhiker Diffuse X-ray Spectrometer and conduct experiments with it in the Shuttle payload bay. The author describes crew members, daily activities, launch preparation, and payloads during the seven-day mission.

Schuiling, Roelof. "STS-55 Mission Report." *Spaceflight*, July 1993. This article discusses STS-55, Space Shuttle *Columbia*'s mission to carry into orbit the reusable German Spacelab D-2 and to use it in the Shuttle's payload bay to conduct 88 experiments in astronomy, atmospheric physics, Earth observations, materials and life sciences, and technology applications. The author describes crew members, daily activities, launch preparation, and payloads during the 11-day mission.

Schuiling, Roelof. "STS-56 Mission Report." *Spaceflight*, June 1993. This article discusses STS-56, Space Shuttle *Discovery*'s mission to carry Atmospheric Laboratory for Applications and Science 2 (ATLAS 2) into orbit and to use it to conduct experiments in the Shuttle payload bay. ATLAS 2 was designed to collect data on the relationship between the Sun's energy output and Earth's middle atmosphere. The author describes crew members, daily activities, launch preparation, and payloads during the 10-day mission.

Schuiling, Roelof. "STS-57 Mission Report." *Spaceflight*, September 1993. This article discusses STS-57, Space Shuttle *Endeavour*'s mission to carry Spacehab into orbit and to use it to conduct biomedical and materials science experiments in the Shuttle payload bay. On this mission, Shuttle astronauts also conducted a spacewalk to retrieve the European Retrievable Carrier and stow it in the Shuttle's payload bay for return to Earth. The author describes crew members, daily activities, launch preparation, and payloads during the 11-day mission.

Schuiling, Roelof. "STS-58 Mission Report." *Spaceflight*, January 1994. This article discusses STS-58, Space Shuttle *Columbia*'s mission to carry into orbit the second dedicated Spacelab for Life Sciences and to use it to conduct experiments in the Shuttle payload bay. Shuttle crew used Spacelab to conduct 14 experiments in cardiovascular and cardiopulmonary physiology, musculoskeletal physiology, neuroscience, and regulatory physiology. The author describes launch crew members, daily activities, launch preparation, and payloads during the 15-day mission.

Schuiling, Roelof. "STS-59 Mission Report." *Spaceflight*, August 1994. This article discusses STS-59, Space Shuttle *Endeavour*'s mission to carry the Space Radar Laboratory (SRL) into orbit and to use it to conduct experiments in the Shuttle payload bay. The SRL, containing instruments such as the Spaceborne Imaging Radar-C, was designed to study

Earth's ecosystem. The author describes crew members, daily activities, launch preparation, and payloads during the 12-day mission.

Schuiling, Roelof. "STS-60 Mission Report." *Spaceflight*, April 1994. This article describes STS-60, the first mission of the U.S.-Russian Shuttle-*Mir* program and the first Space Shuttle flight of a Russian cosmonaut, Sergei K. Krikalev. STS-60's primary mission was to deploy into low Earth orbit the Wake Shield Facility 1 (WSF-1), a free-flying science platform. However, because of technical difficulties, the crew of Space Shuttle *Discovery* was unable to release WSF-1. STS-60's secondary mission was to carry into orbit the commercially developed Spacehab laboratory module and to use it to conduct experiments in the Shuttle payload bay. The author describes crew members, daily activities, launch preparation, and payloads during the nine-day mission.

Schuiling, Roelof. "STS-62 Mission Report." *Spaceflight*, June 1994. This article discusses STS-62, Space Shuttle *Columbia*'s mission to carry into orbit NASA's U.S. Microgravity Payload-2 (USMP-2) and Office of Aeronautics and Space Technology-2 (OAST-2) and to use them to conduct experiments in the Shuttle payload bay. USMP-2 consisted of five experiments investigating materials processing and crystal growth in microgravity. NASA's OAST-2 featured six experiments focusing on space technology and spaceflight. The author describes crew members, daily activities, launch preparation, and payloads during the 15-day mission.

Schuiling, Roelof. "STS-63 Mission Report." *Spaceflight*, May 1995. This article discusses STS-63, the second mission of the U.S.-Russian Shuttle-*Mir* program. In STS-63, *Discovery* carried out the first rendezvous of a U.S. Space Shuttle with Russia's space station *Mir*. Known as the Near-*Mir* mission, STS-63 carried into orbit Spacehab 3 and the Space Radar Laboratory (SRL). STS-63 was the first flight in which a woman—astronaut Eileen M. Collins—piloted the Shuttle, and the second Shuttle flight carrying a Russian cosmonaut—Vladimir G. Titov. The author describes the nine-day mission's crew members, daily activities, launch preparation, and payloads.

Schuiling, Roelof. "STS-64 Mission Report; Laser Atmospheric Research, Robotic Operations, Untethered Spacewalk." *Spaceflight*, December 1994. This article discusses STS-64, Space Shuttle *Discovery*'s mission to conduct atmospheric research with a laser, use a robot to process semiconductor materials, and perform an untethered spacewalk. The author describes preflight processing, crew members, and Shuttle activities for each of the 12 days that the Shuttle was in space.

Schuiling, Roelof. "STS-65 Mission Report." *Spaceflight*, October 1994. This article discusses STS-65, Space Shuttle *Columbia*'s mission to carry into orbit the second flight of the International Microgravity Laboratory 2 (IML-2). The author describes crew members, daily activities, launch preparation, and payloads during the 16-day mission.

Schuiling, Roelof. "STS-66 Mission Report." *Spaceflight*, March 1995. This article discusses STS-66, the mission of Space Shuttle *Atlantis* to carry into orbit seven instruments on the Atmospheric Laboratory for Applications, as well as the Science-3 Cryogenic

Infrared Spectrometers and Telescopes for the Atmosphere-Shuttle Pallet. The author describes crew members, daily activities, launch preparation, and payloads during the 12-day mission.

Schuiling, Roelof. "STS-67 Mission Report." *Spaceflight*, July 1995. This article discusses STS-67, Space Shuttle *Endeavour*'s mission to carry into orbit the ASTRO Observatory and its three ultraviolet telescopes: the Hopkins Ultraviolet Telescope, the Wisconsin Ultraviolet Photo-Polarimeter Experiment, and the Ultraviolet Imaging Telescope. The author describes crew members, daily activities, launch preparation, and payloads during the 17-day mission.

Schuiling, Roelof. "STS-68 Mission Report." *Spaceflight*, February 1995. This article discusses STS-68, Space Shuttle *Endeavour*'s mission to carry into orbit the Space Radar Laboratory (SRL) and to use it to conduct experiments in the Shuttle payload bay. The SRL was designed to help scientists distinguish between human-induced environmental changes and other natural forms of change. STS-68 was the second flight in 1994 of SRL. NASA flew the SRL during different seasons to compare changes between the first and second flights. The author describes STS-68 crew members, daily activities, launch preparation, and payloads during the 12-day mission.

Schuiling, Roelof. "STS-69 Mission Report." *Spaceflight*, December 1995. This article discusses STS-69, Space Shuttle *Endeavour*'s mission to deploy and retrieve Spartan 201-03 and the Wake Shield Facility 2 (WSF-2), an experimental science platform. The author describes crew members, daily activities, launch preparation, and payloads during the 12-day mission.

Schuiling, Roelof. "STS-70 Mission Report." *Spaceflight*, November 1995. This article discusses STS-70, Space Shuttle *Discovery*'s mission to deploy the Tracking and Data Relay Satellite-G (TDRS-G) and to carry into orbit and to conduct experiments such as the Biological Research in Canister, which investigates the effects of spaceflight on small arthropod animal and plant specimens. The author describes crew members, daily activities, launch preparation, and payloads during the 10-day mission.

Schuiling, Roelof. "STS-73 Mission Report." *Spaceflight*, January 1996. This article discusses STS-73, Space Shuttle *Columbia*'s mission to carry the second United States Microgravity Laboratory (USML-2) into space for microgravity studies in the Shuttle payload. The author describes crew members, daily activities, launch preparation, and payloads during the 17-day mission.

Schuiling, Roelof. "STS-77 Mission Report." *Spaceflight*, August 1996. This article discusses STS-77, Space Shuttle *Endeavour*'s mission to carry into orbit Spacehab 4, and to use it to conduct experiments in agriculture, biotechnology, electronic materials, and polymers in the Shuttle payload bay. STS-77 crew also deployed and retrieved the Spartan-207 free flyer. The author describes crew members, daily activities, launch preparation, and payloads during the 11-day mission.

Schuiling, Roelof. "STS-78 Mission Report." *Spaceflight*, October 1996. This article discusses STS-78, Space Shuttle *Columbia*'s mission to carry into space the Life and Microgravity Spacelab, to use it to conduct experiments in the Shuttle payload bay, and to study the effects of long-duration spaceflight on human physiology. The author describes crew members, daily activities, launch preparation, and payloads during the 18-day mission.

Schuiling, Roelof. "STS-80 Mission Report." *Spaceflight*, March 1997. This article discusses STS-80, Space Shuttle *Columbia*'s mission to deploy and retrieve the Orbiting and Retrievable Far and Extreme Ultraviolet Spectrometer-Shuttle Pallet Satellite 2 and the Wake Shield Facility 3 (WSF-3), an experimental science platform. The author describes crew members, daily activities, launch preparation, and payloads during the 18-day mission.

Schuiling, Roelof. "STS-83 Mission Report." *Spaceflight*, October 1997. This article discusses STS-83, Space Shuttle *Columbia*'s mission to carry into orbit the Microgravity Science Laboratory-1 and to use it to conduct 19 materials science investigations in the Shuttle payload bay. The mission was cut short because of concern over erratic readings from some Shuttle fuel cells. The author describes crew members, daily activities, launch preparation, and payloads during the four-day mission.

Schuiling, Roelof. "STS-85 Mission Report." *Spaceflight*, November 1997. This article discusses STS-85, Space Shuttle *Discovery*'s mission to deploy and retrieve the Cryogenic Infrared Spectrometers and Telescopes for the Atmosphere-Shuttle Pallet Satellite-2, as well as to carry into space a number of payloads involving secondary experiments. The author describes crew members, daily activities, launch preparation, and payloads during the 11-day mission.

Schuiling, Roelof. "STS-87 Mission Report." *Spaceflight*, April 1998. This article discusses STS-87, Space Shuttle *Columbia*'s mission to conduct science experiments and to deploy and retrieve the SPARTAN-201-04 free-flyer, a Solar Physics Spacecraft. Experiments conducted in the Shuttle payload bay on the U.S. Microgravity Payload focused on combustion science, fundamental physics, and materials science. The author describes crew members, daily activities, launch preparation, and payloads during the 15-day mission.

Schuiling, Roelof. "STS-90 Mission Report." *Spaceflight*, August 1998. This article discusses STS-90, Space Shuttle *Columbia*'s mission to study neuroscience. The *Columbia* payload included Neurolab and its 26 experiments related to the nervous system. The author describes crew members, daily activities, launch preparation, and payloads during the 15-day mission.

Schuiling, Roelof. "STS-91 Mission Report." *Spaceflight*, November 1998. This article discusses STS-91, Space Shuttle *Discovery*'s mission to dock with the Russian space station *Mir* and to deliver cargo, science experiments, and supplies. In addition, the crew moved long-term U.S. experiments that had been aboard *Mir* into *Discovery*'s mid-deck locker area. The author describes crew members, daily activities, launch preparation, and payloads during the nine-day mission.

Schuiling, Roelof. "STS-93 Launch Delays: Problems on Ascent." *Spaceflight*, December 1999. This article discusses STS-93, Space Shuttle *Columbia*'s mission to deploy the Chandra X-ray Observatory. STS-93 was the first mission in Space Shuttle history to be commanded by a woman—Eileen M. Collins. The author describes crew members, daily activities, launch preparation, and payloads during the six-day mission.

Schuiling, Roelof. "STS-94 Mission Report." *Spaceflight*, October 1997. This article discusses STS-94, Space Shuttle *Columbia*'s repeat mission to carry into orbit the Microgravity Science Laboratory-1 (MSL-1) after erratic fuel cell readings shortened MSL-1's trip aboard STS-83. To prepare for quick relaunch of *Columbia*, NASA serviced MSL-1 in the Shuttle payload bay between STS-83 and STS-94. The author describes crew members, daily activities, launch preparation, and payloads during the 15-day mission.

Schuiling, Roelof. "STS-95: 'The John Glenn Flight'." *Spaceflight*, February 1999. This article discusses STS-95, Space Shuttle *Discovery*'s mission to carry into orbit the pressurized Spacehab module and to use it to conduct a variety of scientific experiments in the Shuttle payload bay. On that mission, Shuttle crew members also deployed and retrieved the Spartan free-flyer payload, which gathered measurements of the solar corona and solar wind. Former Mercury program astronaut and U.S. Senator John H. Glenn Jr. was one of the STS-95 crew members, returning to space at the age of 77. The author describes the 10-day mission's crew members, daily activities, launch preparation, and payloads.

Schuiling, Roelof. "STS-99: Mapping the Earth's Surface by Radar." *Spaceflight*, May 2000. This article discusses STS-99, Space Shuttle *Endeavour*'s mission to map the Earth's surface using radar. The author describes the mission payload, which included the Shuttle Radar Topography Mission, the STS-99 crew, and the crew's activities on each of the 12 days that *Endeavour* was in space.

Schuiling, Roelof. "STS-107: Columbia's Final Mission." *Spaceflight*, April 2003. This article discusses Space Shuttle *Columbia*'s final mission, STS-107, describing launch preparation, payloads, crew members, and daily activities during the 15-day mission to conduct scientific experiments. *Columbia*'s Spacehab RDM held nine commercial payloads involving 21 separate investigations, four payloads for ESA with 14 investigations, one payload/investigation for ISS Risk Mitigation, and 18 payloads supporting 23 investigations for NASA's Office of Biological and Physical Research. *Columbia* and its crew were lost during reentry over east Texas on 1 February 2003, at about 9:00 a.m. (EST).

Simpson, Clive. "STS-114: Return of the Space Shuttle." *Spaceflight*, October 2005. This article discusses STS-114, Space Shuttle *Discovery*'s mission to deliver equipment and supplies to the ISS. The author describes crew members, daily activities, launch preparation, and payloads during the 13-day mission.

Souza, Kenneth, Guy Etheridge, and Paul X. Callahan, eds. *Life into Space: Space Life Sciences Experiments; Ames Research Center, Kennedy Space Center, 1991–1998.* NASA Special Publication 2000-534, Washington, DC, 2000. This book covers various aspects of space life sciences research, including summaries of mission operations, payloads, and experiments developed or managed by NASA's Ames Research Center and NASA's Kennedy Space Center between 1991 and 1998. The book also includes interviews with NASA personnel about various aspects of space life sciences research. Three appendices describe the methods and results of more than 200 flight experiments, the resulting science publications, and associated flight hardware, including illustrations.

Souza, Kenneth, Robert Hogan, and Rodney Ballard, eds. *Life into Space: Space Life Sciences Experiments. NASA Ames Research Center 1965–1990.* NASA Reference Publication 1372, Washington, DC, 1995. This book profiles the study of biological and biomedical processes using live specimens such as microorganisms, cell cultures, plants, and animals aboard Space Shuttle flights. Early experiments focused on the viability of living systems in the microgravity environment. Later, changes that occur in living systems in response to microgravity were studied. More recently, experiments focused on attempts to understand the mechanisms for changes observed, and to develop methods to oppose those changes. The book contains one-page summaries of more than 200 completed experiments.

"Spacelab Experiments." *Spaceflight*, October 1996. This article gives short descriptions of Spacelab experiments during STS-78. There is information about human physiology; plant and animal; materials science; and microgravity experiments.

The Space Shuttle. Bethesda, MD: Discovery Communications, 1996. Videocassette (VHS). This video, intended for grades 6–12, shows how NASA prepares the Space Shuttle for liftoff, as well as how on-board experiments open new worlds of opportunity for scientific exploration.

U.S. General Accounting Office. "Space Station: Plans To Expand Research Community Do Not Match Available Resources." Report no. GAO/NSIAD-95-33, Washington, DC, November 1994. *http://www.gao.gov/archive/1995/ns95033.pdf* (accessed 8 March 2012). This report focuses on NASA's Earth-orbiting microgravity and life sciences research laboratory. The GAO reports that NASA is focusing on developing a comprehensive research program that emphasizes more ground-based research and uses spaceflight only for research efforts that require a microgravity environment in space.

Vastag, Brian. "Shuttle Packs $2 Billion Physics Experiment." *Washington Post*, 29 April 2011. During STS-134, Space Shuttle *Endeavour* carried the Alpha Magnetic Spectrometer (AMS) to the ISS. The AMS, a US$2 billion, 7-ton (6.4-tonne or 6,350-kilogram) experiment, will sniff space for cosmic rays, antimatter, dark matter, and other exotic and poorly understood phenomena.

Wilson, Keith T. "From Landing . . . To Launch." *Spaceflight*, October 1997. This article describes the short turnaround time (94 days) between Space Shuttle *Columbia*'s STS-83 and STS-94 missions, as well as the missions' prelaunch processing. The staff of NASA's Kennedy Space Center staff tested and evaluated the Shuttle for its return to space. The Microgravity Science Laboratory remained in the Shuttle between the two flights, the first time that a major payload had remained on-board while the Shuttle was on the ground.

Link to Part 1 (1970–1991), Chapter 9—Science on the Shuttle, Potential and Actual

CHAPTER 9—COMMERCIAL USES OF THE SPACE SHUTTLE

Anselmo, Joseph C. "NASA To Seek Major Shift in U.S. Shuttle Policy." *Aviation Week & Space Technology*, 13 October 1997. This article describes NASA's support for United Space Alliance's lobby to rescind an 11-year-old presidential edict and a provision in U.S. law prohibiting the Space Shuttle from carrying commercial satellites into orbit. United Space Alliance, the venture that manages Shuttle operations, is promoting these policy changes in the hope of completely privatizing Shuttle operations within five years. The expendable launch vehicle industry is likely to oppose the move so that it can retain its monopoly on commercial spacecraft launches.

Calvert, Ken. "Exploration Needs Commercial Space Transportation." *Aviation Week & Space Technology*, 20 February 2005. In this article, U.S. House Representative Kenneth S. Calvert (R-CA), chair of the House Science Committee's Space and Aeronautics Subcommittee, calls for increased commercialization of the space program and a regulatory framework that allows private human spaceflight.

Covault, Craig. "Shuttle Privatization Raises Safety Issues." *Aviation Week & Space Technology*, 24–31 December 2001. NASA's Johnson Space Center has finished an initial privatization assessment involving a large team from NASA's field centers, United Space Alliance, and other Shuttle contractors. The report concludes that continued dependence on support from NASA's civil service poses a greater risk to Shuttle safety than privatization.

Covault, Craig. "Station Commercialization Set as Assembly Flight Readied." *Aviation Week & Space Technology*, 30 November 1998. This article describes Shuttle *Endeavour*'s STS-88, the first Space Shuttle mission to assemble the initial elements of the ISS, and discusses NASA's plan to turn the multibillion-dollar ISS into a commercial, fee-for-service facility. According to this scenario, private industry would assume substantial responsibility for the space station.

Covault, Craig. "Unusual Shuttle Operations Advance Commercialization." *Aviation Week & Space Technology*, 2 December 1996. This article describes STS-80 and the difficulty the *Columbia* crew had in releasing the 2-ton (1.8-tonne) commercial Wake Shield Facility (WSF), an experimental science platform, from the Shuttle payload. The WSF, used to produce wafer semiconductors in space, is an example of the Space Shuttle's commercial utility.

Figliola, Patricia Moloney, Carl E. Behrens, and Daniel Morgan. "U.S. Space Programs: Civilian, Military, and Commercial." CRS Issue Brief for Congress, Congressional Research Service, Library of Congress, Washington, DC, 13 June 2006. *http://digital. library.unt.edu/ark:/67531/metacrs10507/?q=IB92011* (accessed 8 March 2012). This report discusses how to manage DOD space programs, avoiding the cost growth and schedule delays that have characterized several recent projects. The authors also describe the appropriate role of the government in facilitating commercial space businesses.

Griffin, Gerry. "As Shuttle Retires, a Vote for Commercial Space Flight." *USA Today*, 6 April 2011. The author reports that, because it will not have low-Earth-orbit transportation capability after the Space Shuttle program ends, NASA will not be able to explore and learn more about space. Furthermore, NASA will not be able to conduct unpiloted spaceflight once the Space Shuttle retires. However, the author believes that the commercial spaceflight industry shows encouraging signs that it may develop the capability of conducting human spaceflight in the near future.

Morring, Frank, Jr. "Commercial Break; NASA Plans COTS-Only Approach for ISS, Dropping Russia's Progress." *Aviation Week & Space Technology*, 21 April 2008. This article reports that NASA officials will discuss with Congress a plan for NASA to continue using Russia's Soyuz crew launch vehicles to transport astronauts to and from the ISS after the final Space Shuttle flight in 2010. NASA does not intend to continue using Russian Progress vehicles for U.S. cargo resupply but plans to use its own Commercial Orbital Transportation System (COTS) program vehicles, which are as yet untested.

Shiga, David. "Cutting the Cost of Spaceflight; NASA Is Looking to Commercial Spacecraft to Lower the Cost of Ferrying Cargo and Crew to the International Space Station." *New Scientist*, 27 September 2008. This article provides an overview of how the Space Shuttle was developed and reports on the commercial space race. The author claims that the spacecraft Dream Chaser, developed by SpaceDev Company, may lower the cost of access to space.

Smith, Marcia S. "Space Launch Vehicles: Government Activities, Commercial Competition, and Satellite Exports." CRS Brief for Congress, Congressional Research Service, Library of Congress, Washington, DC, 31 January 2006. *http://digital.library.unt.edu/ark:/67531/ metacrs10160/m1/1/high_res_d/IB93062_2006Jan31.pdf* (accessed 8 March 2012). This report describes U.S. launch vehicle policies, programs, and issues; the U.S. commercial launch service industry; foreign launch competition; and satellite exports.

The Space Shuttle. DVD. New York: Jaffe Productions, Hearst Entertainment, A & E Home Video, New Video Group, and History Channel, 2004. This DVD program details the development of the Space Shuttle from the 1950s to its triumphant launch in 1981. The program examines the successes and failures of the Shuttle missions and looks at the next generation Shuttle—the futuristic commercial reusable space vehicle X-33 VentureStar.

U.S. Congress. House of Representatives. Committee on Science. Subcommittee on Space and Aeronautics. *Space Shuttle and Space Launch Initiative*. 107th Cong., 2nd sess., 18 April 2002. *https://web.lexis-nexis.com/congcomp/* (accessed 11 November 2011 via Proquest Congressional). This hearing reviews proposed Space Shuttle safety and performance upgrades and examines the NASA Space Launch Initiative, a program for research into the development and commercial applications of advanced and alternative space transportation technologies, including RLV development.

U.S. Congress. House of Representatives. Committee on Science. Subcommittee on Space and Aeronautics. *Space Transportation, Parts I–IV*. 106th Cong., 1st sess., 29 September;

13–27 October 1999. *https://web.lexis-nexis.com/congcomp/* (accessed 11 November 2011 via Proquest Congressional). This hearing reviews the status of NASA's RLV program, including the X-33 RLV demonstration program; assesses private-sector efforts to develop RLVs using private capital; evaluates proposed safety and performance upgrades to the Space Shuttle; and examines the development of future space transportation systems.

U.S. Congress. Senate. Committee on Commerce, Science, and Transportation. Subcommittee on Science and Space. *Assessing Commercial Space Capabilities.* 111th Cong., 2nd sess., 18 March 2010. *http://frwebgate.access.gpo.gov/cgi-bin/getdoc.cgi?dbname=111_senate_ hearings&docid=f:66983.pdf* (accessed 14 March 2012). This hearing examines commercial space capabilities and developments in light of the decision of the George W. Bush administration to discontinue the Space Shuttle program in favor of expanded commercial sector contracting, under NASA oversight, for transportation of astronauts to the ISS and for further exploration.

U.S. General Accounting Office. "Space Transportation: Status of the X-33 Reusable Launch Vehicle Program." Report no. GAO/T-NSIAD-99-243, Washington, DC, 29 September 1999. *http://www.gao.gov/archive/1999/ns99243t.pdf* (accessed 8 March 2012). The report focuses on the possible phaseout of the Space Shuttle and its replacement with commercial launch services. The GAO remarks that the program will not meet some of its original cost, schedule, and performance objectives because of problems developing technologies for the X-33 VentureStar. In addition, the report states that the X-33 will not carry as much cargo as the Space Shuttle and will have to dock at the ISS more frequently than the Shuttle does.

Williamson, Ray A. "The US-Europe Technology Gap in Space Transportation: The View from the USA." *Space Policy* 17, no. 1 (February 2001): 27–33. This article reports that competition with Europe's Ariane launcher has influenced the U.S. decision to privatize the expendable launch vehicle production and operation industry. The author also reports that the U.S. government has decided to fly all government payloads on the Space Shuttle and to market Space Shuttle payloads to the private industry.

Link to Part 1 (1970-1991), Chapter 10—Commercial Uses of the Shuttle

CHAPTER 10—THE SPACE SHUTTLE AND THE MILITARY

Cassutt, Michael. "Secret Space Shuttles." *Air & Space*, August 2009. This article discusses several classified U.S. Space Shuttle missions conducted jointly by NASA, the U.S. Air Force, and the intelligence community during the 1980s. One of the missions, STS-27, deployed the first of a series of spy satellites that use radar to observe ground targets, day or night and in any kind of weather.

Day, Dwayne A. "Department of Defense Returns to Space Shuttle." *Spaceflight*, February 1997. This article describes a cooperative agreement between NASA and the National Imagery and Mapping Agency to fly a joint mission. The Shuttle Radar Topography Mission will attempt to make an accurate map of the Earth using SIR-C radar.

Day, Dwayne A. "Out of the Shadows: The Shuttle's Secret Payloads." *Spaceflight*, February 1999. This article describes secret payloads launched by NASA on the Space Shuttle under agreements with the National Reconnaissance Office. The article describes the satellites carried in the payloads and the orbits into which the satellites were deployed.

Day, Dwayne A. "Secret Shuttle Mission Revealed." *Spaceflight*, July 1998. This article describes the military payload aboard STS-51-J. During Space Shuttle *Atlantis*'s first orbit, the military crew deployed two Defense Satellite Communications Systems-3 satellites. The article describes the mission and the satellites.

Figliola, Patricia Moloney, Carl E. Behrens, and Daniel Morgan. "U.S. Space Programs: Civilian, Military, and Commercial." CRS Issue Brief for Congress, Congressional Research Service, Library of Congress, Washington, DC, 13 June 2006. *http://digital. library.unt.edu/ark:/67531/metacrs10507/?q=IB92011* (accessed 8 March 2012). This report discusses how to manage DOD space programs, avoiding the cost growth and schedule delays that have characterized several recent projects. The authors also describe the appropriate role of the government in facilitating commercial space businesses.

Moltz, James Clay. *The Politics of Space Security: Strategic Restraint and the Pursuit of National Interests*. Palo Alto, CA: Stanford University Press, 2008. This book describes 50 years of space security. The author recounts trends in military space developments and argues that the United States and the former Soviet Union—now Russia—have shown restraint in militarizing space in order to protect access to critical military and civilian assets in orbit. Although the book focuses on space weapons, the author suggests that the Space Shuttle could capture a small spacecraft in orbit, such as a satellite.

Powell, Joel W. "Secret Shuttle Payloads Revealed." *Spaceflight*, May 1993. This article describes the DOD payloads aboard STS-27, STS-28, STS-33, STS-36, STS-38, and STS-51. The article describes military Manned Spaceflight Engineers, military personnel who serve as astronauts and are trained as Shuttle payload specialists.

Schuiling, Roelof. "STS-53 Mission Report." *Spaceflight*, March 1993. This article discusses STS-53, Space Shuttle *Discovery*'s mission to deploy a classified DOD satellite and to conduct two unclassified secondary experiments in the Shuttle payload bay. The author describes crew members, daily activities, launch preparation, and payloads during the eight-day mission.

U.S. General Accounting Office. "Military Space Operations: Common Problems and Their Effects on Satellite and Related Acquisitions." Report no. GAO-03-825R, Washington, DC, 2 June 2003. *http://www.gao.gov/new.items/d03825r.pdf* (accessed 8 March 2012). This report briefly describes the U.S. military's use of the Space Shuttle to launch GPS satellites and other space-related systems, such as a strategic surveillance and warning satellite system with an infrared capability to detect ballistic-missile launches.

U.S. General Accounting Office. "Space Station: Information on National Security Applications and Cost." Report no. GAO/NSIAD-93-208, Washington, DC, May 1993. *http://archive. gao.gov/t2pbat5/149216.pdf* (accessed 8 March 2012). In this report, the GAO explains that the Pentagon has no real need for a piloted space station and could carry out military research using the Space Shuttle. The GAO describes the cost of Space Station Freedom, a NASA project to construct a permanently piloted Earth-orbiting space station in the 1980s. Although approved by President Ronald W. Reagan and announced in the 1984 State of the Union Address, the proposed Space Station Freedom was never constructed or completed as originally designed. After several cutbacks, the remnants of the project became part of the ISS.

Vis, Bert. "The NEREUS Programme." *Spaceflight*, October 1998. This article discusses oceanographers who have ridden on the Space Shuttle for the Navy Environmental Research Experiment Using Shuttle project. Oceanographers flew aboard STS-41G and STS-61K, using special cameras to observe Earth and the oceans.

Link to Part 1 (1970–1991), Chapter 11—The Shuttle and the Military

CHAPTER 11—SPACE SHUTTLE ASTRONAUTS

Adamson, Heather. *The Challenger Explosion*. Mankato, MN: Capstone Press, 2006. Ages 9–12. This graphic novel tells the story of Christa McAuliffe and the six other NASA astronauts who lost their lives in the Space Shuttle *Challenger* disaster on 28 January 1986.

Begley, Sharon, Ginny Carroll, Peter Katel, Catharine Skipp, and Peter Annin. "Down to Earth." *Newsweek*, 7 October 1996. This article profiles astronaut Shannon W. Lucid, examining her record 188 days in space (179 days aboard *Mir* and nine days total on Space Shuttle trips to and from *Mir*) and her health after returning to Earth in September 1996. The authors also discuss Lucid's education and her career background.

Bredeson, Carmen. *Shannon Lucid Space Ambassador*. Brookfield, CT: Millbrook Press, 1998. *http://www.barnesandnoble.com/s/Shannon-Lucid-Space-Ambassador-?keyword= Shannon+Lucid+Space+Ambassador.&store=ebook* (accessed 8 March 2012). This electronic book chronicles the life of astronaut Shannon W. Lucid, from her childhood in Oklahoma, through her various Space Shuttle missions, to her six months aboard the *Mir* space station.

Burgess, Colin. *Australia's Astronauts: Countdown to a Spaceflight Dream*. Berowra, NSW: The Communications Agency, 2009. This book tells the stories of the three Australian astronauts, Philip K. Chapman, Paul D. Scully-Power, and Andrew S. W. Thomas, recounting their experiences and explaining their roles in the space programs of the United States and Russia. Chapman did not fly aboard a Space Shuttle but was the first Australian-born person selected as a scientist-astronaut by NASA. Scully-Power flew on STS-41-G as a payload specialist and was the first Australian-born person to journey into space. Thomas flew aboard STS-77 as a payload commander and aboard STS-77 as a mission specialist. Thomas also trained at the Gagarin Cosmonaut Training Center in Star City, Russia. The book also discusses missed opportunities for other Australian astronauts to fly in space as part of NASA's international payload specialist program.

Chien, Philip. *Columbia—Final Voyage: The Last Flight of NASA's First Space Shuttle*. Chichester, UK: Praxis Publishing, 2006. This book explains STS-107, *Columbia*'s final mission, a "free flyer" mission in which NASA planned for the crew to spend 16 days in orbit, performing dozens of scientific experiments. The book devotes one chapter to each STS-107 astronaut. The author criticizes the media for covering the mission only after the catastrophe had occurred.

Cole, Michael D. *The Columbia Space Shuttle Disaster: From First Liftoff to Tragic Final Flight*. Rev. ed. Berkeley Heights, NJ: Enslow Publishers, 2003. This book contains technical details about the Space Shuttle, as well as personal information about the astronauts aboard *Columbia* on its final mission, STS-107. Photographs introduce young readers to the crew and help them understand the events that led to the tragic loss of the spacecraft and its crew.

Covault, Craig. "Station Training Focus: 'Building on the Fly'." *Aviation Week & Space Technology*, 2 September 1996. This article describes astronaut and cosmonaut training for those who will build and live in the ISS. The author presents information about the Space Shuttle Training Facility at NASA's Johnson Space Center in Houston, Texas, and the Neutral Buoyancy Laboratory underwater training facility. He also explains EVA training requirements. In addition, the article discusses the costs of the ISS and its assembly schedule.

DiGregorio, Barry. "The Right Stuff; Grounded for More Than Two Years After the Loss of Columbia, the Space Shuttle Is Due To Take to the Skies Again Sometime in July." *New Scientist*, 14 May 2005. This article presents an interview with Eileen M. Collins, Commander of *Discovery* for the first flight of the Space Shuttle since *Columbia*'s accident. Collins is confident that the redesign of the Shuttle's external fuel tank and the removal of large pieces of foam from the front section of the tank will protect *Discovery* from the type of accident that caused the loss of *Columbia*.

Foale, Colin. *Waystation to the Stars: The Story of Mir, Michael, and Me*. London: Headline Book Publishing, 2000. This book, written by the father of British-American astronaut C. Michael Foale, describes the astronaut's activities and experiences on *Mir* for five months in 1997. During his stay on *Mir*, Foale had to deal with fire, collision, and computer failure.

French, Francis. "The Unassuming Journey of Charlie Bolden." *Spaceflight*, March 2002. This article recounts how current NASA Administrator Charles F. Bolden Jr. overcame personal obstacles to attend the U.S. Naval Academy and become a U.S. Marine Corps aviator before joining the astronaut program and becoming a Space Shuttle astronaut.

Harris, Bernard A., Jr. *Dream Walker: A Journey of Achievement and Inspiration*. Austin, TX: Greenleaf Book Press Group, 2010. This book by Bernard A. Harris Jr., Mission Specialist on STS-55 and Payload Commander on STS-63, describes Harris's modest background and his experiences in college, medical school, and during training as a NASA flight surgeon. Harris was the first African American to walk in space.

Hawaleshka, Danylo. "A Hero in Orbit: John Glenn's Return Sparks Interest in the Space Saga." *Maclean*'s, 2 November 1998. This article examines reactions to the 77-year-old John H. Glenn Jr.'s inclusion in a Space Shuttle *Discovery* mission, STS-95. In 1962 Glenn became the first American to orbit the Earth. Although critics claim that Glenn's return to space on board the Space Shuttle is simply a publicity stunt, NASA asserts that Glenn's participation in the flight is a contribution to science. Monitoring the oldest man ever to fly in space will help NASA investigate the detrimental effects of microgravity on astronauts and could also help improve health care for the elderly.

Holden, Henry M. *The Tragedy of the Space Shuttle Challenger*. Berkeley Heights, NJ: MyReportLinks.com Books, 2004. Ages 9–12. This book describes events surrounding the explosion of Space Shuttle *Challenger* in 1986 and discusses the investigation of this disaster. The author tells the stories of the seven astronauts who died. The book also includes Internet links to related Web sites, source documents, and photographs.

Jones, Thomas D. "The Future of NASA's Astronaut Corps." *Aerospace America*, October 2010. In this article, a former astronaut compares the number of astronauts at NASA when he flew his first Shuttle flight in 1994 with the number of astronauts that NASA will need after the Space Shuttles retire. Jones also discusses astronaut hiring and training for ISS missions.

Kevles, Bettyann Holtzmann. *Almost Heaven: The Story of Women in Space*. New York, NY: Basic Books, 2003. This book tells the stories of the first women to face the risks of space travel, discussing their contributions to science and to society. The author traces the careers of several female astronauts and cosmonauts who broke new trails in a traditionally male profession. The publisher issued an updated paperback edition in 2006.

Kosova, Weston, Sam Seibert, Seth Mnookin, and Joshua Hammer. "The Right Stuff." *Newsweek*, 10 February 2003. This article profiles the astronauts who perished on *Columbia*: Richard D. Husband, William C. McCool, Michael P. Anderson, Kalpana Chawla, David M. Brown, Laurel B. Clark, and Ilan Ramon.

Lenehan, Anne E. *Story: The Way of Water*. Berowra, NSW: The Communications Agency, 2004. This book is about astronaut Franklin Story Musgrave, who flew aboard STS-6, STS-51, STS-33, STS-44 STS-61, and STS-80. Musgrave is a doctor, space physiologist, pilot, astronaut, poet, philosopher, and artist.

Lieurance, Suzanne. *The Space Shuttle Challenger Disaster in American History*. Berkeley Heights, NJ: Enslow Publishers, 2001. Ages 9–12. Grades 5–7. After recounting the events of the morning that Space Shuttle *Challenger* exploded, the author places the *Challenger*'s mission in historical context. She summarizes the U.S. space program up to that time, profiles each of the astronauts assigned to the *Challenger* crew and discusses the selection and training of civilian teacher Christa McAuliffe for the mission.

Linenger, Jerry M. *Off the Planet: Surviving Five Perilous Months Aboard the Space Station Mir*. New York: McGraw-Hill, 2000. In this book, astronaut Jerry M. Linenger describes his background and his experience training for and living aboard *Mir*. Linenger recounts his efforts to survive 132 days aboard the decaying and unstable Russian space station *Mir*.

McNair, Carl S., and H. Michael Brewer. *In the Spirit of Ronald E. McNair, Astronaut: An American Hero*. Atlanta, GA: Publishing Associates, 2005. This book describes the life and career of Ronald E. McNair, who died in STS-51's *Challenger* tragedy. From a modest background, McNair graduated from college and earned a PhD in physics, becoming one of the world's first African American Space Shuttle astronauts. McNair flew on missions STS-41 and STS-51.

Melady, John. *Canadians in Space: The Forever Frontier*. Toronto: Dundurn Press, 2009. This book, commemorating the 25th anniversary of the flight of the first Canadian astronaut, Marc Garneau, focuses on the eight Canadian astronauts who have flown on the Space Shuttle.

Mullane, R. Mike. *Liftoff! An Astronaut's Dream*. Parsippany, NJ: Silver Burdett, 1995. The author, a former astronaut, describes his experiences in space and shares his ideas about the future of spaceflight.

Musgrave, Story, Lance Lenehan, and Anne Lenehan. *The NASA Northrop T-38: Photographic Art from an Astronaut Pilot*. Kissimmee, FL: Lannistoria, 2008. For decades, Space Shuttle astronauts have trained in the Northrop T-38 Talon, a twin-engine supersonic jet. In this book, astronaut Franklin Story Musgrave shares with space enthusiasts and photographers his private collection of photographs of the Northrop T-38 Talon. Accompanying the images, Musgrave's first-hand account of people and airplanes provides a fascinating insight into the culture of NASA's flying world.

NASA. Johnson Space Center. *STS-107 Memories*. Washington, DC: U.S. Government Printing Office, 2006. CD. This CD, commemorating the seven crew members who lost their lives on Space Shuttle *Columbia* in February 2003, features biographies and photographs of the crew as they worked and trained together.

Reichhardt, Tony. "Shuttlenauts: The Faces of the Space Shuttle Era." *Air & Space*, January 2011. The author discusses the Shuttle's role in the construction of the ISS and examines the careers of several astronauts who spent time on the Space Shuttle, including former Navy pilot Robert L. Crippen and Shuttle Commander Peggy A. Whitson.

Reichhardt, Tony, ed. *Space Shuttle: The First 20 Years—The Astronauts' Experiences in Their Own Words*. New York: Dorling Kindersley, 2002. This book, compiled by the editors of *Air & Space* and *Smithsonian* magazines, documents the history of the Space Shuttle program based on astronauts' anecdotes and reminiscences. The book includes 77 first-person accounts, including astronauts' descriptions of their experiences in zero gravity and their fear of failing in their missions.

Rumerman, Judy A. *Human Space Flight: A Record of Achievement, 1961–1998*. Monographs in Aerospace History no. 9, NASA History Division, Office of Policy and Plans, NASA Headquarters, Washington, DC, August 1998. *http://history.nasa.gov/40thann/humanspf. htm* (accessed 8 March 2012). This book describes the United States' record of achievement in human spaceflight, from the *Mercury* experimental flights, to the *Apollo* Moon landings, to the current flights of the Space Shuttle. The author provides brief mission descriptions and a list of crew members for Shuttle flights STS-1 through STS-94.

Saslow, Rachel. "A Man Who Spent Years at NASA and 35 Days in Space." *Washington Post*, 5 July 2011. The author interviews former Space Shuttle astronaut Piers J. Sellers about his memories of spaceflight and the end of the U.S. Space Shuttle program.

Schefter, Jim. "The Right Stuff-Again." *Popular Science*, May 1998. This article profiles U.S. Senator John H. Glenn Jr., the first American astronaut to orbit the Earth in the 1960s, discussing his preparations for his 1998 mission on the Space Shuttle.

Schuiling, Roelof. "STS-63 Mission Report." *Spaceflight*, May 1995. This article discusses STS-63, the second mission of the U.S.-Russian Shuttle-*Mir* program. In STS-63, *Discovery* carried out the first rendezvous of a U.S. Space Shuttle with Russia's space station *Mir*. Known as the Near-*Mir* mission, STS-63 carried into orbit Spacehab 3 and the Space Radar Laboratory (SRL). STS-63 was the first flight in which a woman—astronaut Eileen M. Collins—piloted the Shuttle, and the second Shuttle flight carrying a Russian cosmonaut—Vladimir G. Titov. The author describes the nine-day mission's crew members, daily activities, launch preparation, and payloads.

Schuiling, Roelof. "STS-95: 'The John Glenn Flight'." *Spaceflight*, February 1999. This article discusses STS-95, Space Shuttle *Discovery*'s mission to carry into orbit the pressurized Spacehab module and to use it to conduct a variety of scientific experiments in the Shuttle payload bay. On that mission, Shuttle crew members also deployed and retrieved the Spartan free-flyer payload, which gathered measurements of the solar corona and solar wind. Former Mercury program astronaut and U.S. Senator John H. Glenn Jr. was one of the STS-95 crew members, returning to space at the age of 77. The author describes the 10-day mission's crew members, daily activities, launch preparation, and payloads.

Schwartz, John. "Astronaut Profiles." *New York Times*, 7 August 2007. This article profiles the seven members of the crew of Space Shuttle *Endeavour* for STS-118, which launched on 8 August 2007.

Schwartz, John. "The Astronauts of STS-120." *New York Times*, 23 October 2008. This article profiles crew members of STS-120, who launched aboard Shuttle *Discovery* on 23 October 2007.

Stott, Carole. *Fly the Space Shuttle*. New York: Dorling Kindersley, 2001. Ages 9 and older. In this book, aimed at a juvenile audience, the reader is invited to pretend that he or she is training as an astronaut to fly on a Space Shuttle mission. The reader is introduced to other members of the crew and learns about the fun and the difficulties of living in a weightless environment, maneuvering the Shuttle orbiter in space, and deploying a satellite. The book includes an easy-to-make, three-dimensional Space Shuttle model. Pull-out and lift-the-flap sections show how the Shuttle works and how it is used.

"STS-107 Crew Profiles." *Spaceflight*, April 2003. This one-page article describes the seven crew members of STS-107, who perished in the *Columbia* tragedy, providing brief biographical information and a picture of each crew member.

Sweetman, Bill. "Seen from Washington." *Interavia Business & Technology*, May 1999. This article focuses on developments related to aerospace industries in the United States as of May 1999, including NASA's plan to launch a Space Shuttle mission with a crew of women only. The author takes a negative view of an all-female Space Shuttle crew.

U.S. General Accounting Office. "Astronaut Utilization." Report no. GAO/NSIAD-93-114R, Washington, DC, 12 January 1993. *http://archive.gao.gov/d36t11/148414.pdf* (accessed 8 March 2012). In this report, the GAO finds that NASA has no written policies or procedures to determine the size of the astronaut corps and that NASA's flight rates have been lower than anticipated, decreasing the need for pilots. Furthermore, the report states that the nature of space missions has changed, from simple satellite deployments to a variety of space-based activities, increasing the need for mission specialists.

Van Den Berg, Anne. "Story Musgrave." *Spaceflight*, November 1996. This article profiles NASA astronaut Franklin Story Musgrave, Mission Specialist for STS-80. When that mission is complete, Musgrave will have flown on all five Space Shuttles and will also have flown on more Shuttle flights than any other astronaut.

Vogt, Gregory. *John Glenn's Return to Space*. Brookfield, CT: Millbrook Press, 2000. Grades 5–8. This book describes John H. Glenn Jr.'s early days as an astronaut and tells the story of his 1962 spaceflight in Friendship 7, as the first American to orbit Earth. The author recounts Glenn's return to space in 1998 at the age of 77 aboard Space Shuttle *Discovery*, emphasizing the extraordinary contrast between Glenn's first orbit around Earth and his recent Shuttle flight and using Glenn's story to symbolize the evolution of the American space program. The author also explains that Glenn joined the *Discovery* crew as a senior citizen so that he could participate in experiments aimed at measuring the effects of space travel on the human body.

Whitehouse, Patricia. *Living in Space*. Chicago: Heinemann Library, 2004. Ages 4–8. Grades 1–3. This book gives readers a glimpse of the experiences astronauts have while living in space and on the Space Shuttle. It describes Shuttle astronauts' daily routines, and explains how they receive air, water, and food, and reveals what happens to their garbage.

Whitehouse, Patricia. *Working in Space*. Chicago, IL: Heinemann Library, 2003. Ages 6 and up. This introduction to working in space discusses Space Shuttle astronaut training and different kinds of jobs in space, describing what it is like to leave Earth aboard a Space Shuttle, work in microgravity, fix satellites, work outside a Space Shuttle, and put on a spacesuit. The author also discusses what it might be like to work on other worlds. The book includes a section of space facts.

Wilson, Scott. "Obama Picks Shuttle Veteran To Be First Black NASA Chief." *Washington Post*, 24 May 2009. The article announces that President Barack H. Obama has nominated former Marine aviator and Space Shuttle astronaut Charles F. Bolden Jr. to head NASA. As NASA's first African American Administrator, Bolden will oversee a broad review of NASA's ambitions for piloted and robotic space exploration.

Woodmansee, Laura S. *Women Astronauts*. Burlington, ON: Apogee Books, 2002. This book features stories and interviews with many past and current female astronauts, including Shuttle astronauts Ellen S. Baker, Kalpana Chawla, Eileen M. Collins, Bonnie J. Dunbar, Anna L. Fisher, Linda M. Godwin, Ellen Ochoa, and Heidemarie M. Stefanyshyn-Piper.

The book describes their childhoods, training, everyday lives, and missions, explaining the determination, commitment, and expertise required to work in the space industry.

Link to Part 1 (1970–1991), Chapter 12—Shuttle Astronauts

CHAPTER 12—THE SPACE SHUTTLE AND INTERNATIONAL RELATIONS

Behrens, Carl, and Mary Beth Nitikin. "Extending NASA's Exemption from the Iran, North Korea, and Syria Nonproliferation Act." CRS Report for Congress, Congressional Research Service, Library of Congress, Washington, DC, 1 October 2008. *http://assets.opencrs.com/rpts/RL34477_20081001.pdf* (accessed 8 March 2012). This report discusses the Iran Nonproliferation Act of 2000, enacted to prevent foreign transfers to Iran of weapons of mass destruction, missile technology, and advanced conventional weapons technology, particularly transfers from Russia. Section 6 of the Act bans U.S. payments to Russia in connection with the ISS unless the President of the United States determines that Russia is taking steps to prevent the proliferation of weapons to Iran.

Burgess, Colin. *Australia's Astronauts: Countdown to a Spaceflight Dream*. Berowra, NSW: The Communications Agency, 2009. This book tells the stories of the three Australian astronauts, Philip K. Chapman, Paul D. Scully-Power, and Andrew S. W. Thomas, recounting their experiences and explaining their roles in the space programs of the United States and Russia. The book also discusses missed opportunities for other Australian astronauts to fly in space as part of NASA's international payload specialist program.

Burgess, Colin. "Lost Mission." *Spaceflight*, June 1999. This article describes how the loss of Space Shuttle *Challenger* in 1986 affected the future missions of Indonesian astronaut candidates for Space Shuttle missions. The U.S. government offered to train an Indonesian astronaut who would be a crew member on a mission to launch an Indonesian satellite. Post-tragedy launch delays forced NASA to launch the Indonesian satellite on an unmanned booster so the Indonesian astronaut candidates did not get to fly a Shuttle mission. The article also portrays the two astronaut candidates chosen for the mission, Taufik Akbar and Pratiwi Sudarmono.

Fischer, Hans-Jurgen, and German A. Zoeschinger. "Planning and Co-ordination of Space Shuttle Attitudes and Trajectory for Spacelab Mission D-2." *Journal of the British Interplanetary Society* 47, no. 7 (July 1994): 260–265. The article describes how the German Space Operations Center helped plan and coordinate a Space Shuttle payload.

Melady, John. *Canadians in Space: The Forever Frontier*. Toronto: Dundurn Press, 2009. This book, commemorating the twenty-fifth anniversary of the flight of the first Canadian astronaut, Marc Garneau, focuses on the eight Canadian astronauts who have flown on the Space Shuttle.

Pletser, Vladimir. "ESA's Fluid Physics Experiments on STS-78." *Spaceflight*, October 1996. This article describes the fluid physics experiments that ESA conducted aboard *Columbia* during STS-78 in 1996.

Rouss, Sylvia A. *Reach for the Stars: A Little Torah's Journey*. New York: Devora Publishing, 2004. Grades 3–5. This is a true story of the miniature Torah that Israeli astronaut Ilan Ramon carried with him on Space Shuttle *Columbia*'s mission, STS-107. This unique

Torah survived the Holocaust, along with its guardian, Joachim Joseph. Years later, Ilan Ramon became close friends with Joachim Joseph and agreed to take the Torah with him on his journey into space.

Sadeh, Eligar, James P. Lester, and Willy Z. Sadeh. "Modeling International Cooperation for Space Exploration." *Space Policy* 12, no. 2 (August 1996): 207–223. Based on their belief that international cooperation is a prerequisite for space exploration in the twenty-first century, the authors propose a theoretical framework for international cooperation among economic, political, scientific, and technological entities.

Schuiling, Roelof. "STS-46 Mission Report." *Spaceflight*, January 1993. This article discusses STS-46, the mission of Space Shuttle *Atlantis* to deploy ESA's European Retrievable Carrier and to operate the joint project of NASA and Agenzia Spaziale Italiana (the Italian space agency)—the Tethered Satellite System. The author describes launch preparation, payloads, crew members, and activities during the eight-day mission.

Schuiling, Roelof. "STS-47 Mission Report." *Spaceflight*, April 1993. This article discusses STS-47, Space Shuttle *Endeavour*'s mission to carry Spacelab-J into orbit and to use it to conduct experiments in the Shuttle payload. A joint mission of NASA and Japan's space agency, NASDA, Spacelab-J uses an unpiloted Spacelab module to conduct microgravity investigations in materials and life sciences. The author describes launch preparation, payloads, crew members, and daily activities during the eight-day mission.

Schuiling, Roelof. "STS-55 Mission Report." *Spaceflight*, July 1993. This article discusses STS-55, Space Shuttle *Columbia*'s mission to carry into orbit the reusable German Spacelab D-2 and to use it in the Shuttle's payload bay to conduct 88 experiments in astronomy, atmospheric physics, Earth observations, materials and life sciences, and technology applications. The author describes crew members, daily activities, launch preparation, and payloads during the 11-day mission.

Schuiling, Roelof. "STS-57 Mission Report." *Spaceflight*, September 1993. This article discusses STS-57, Space Shuttle *Endeavour*'s mission to carry Spacehab into orbit and to use it to conduct biomedical and materials science experiments in the Shuttle payload bay. On this mission, Shuttle astronauts also conducted a spacewalk to retrieve the European Retrievable Carrier and stow it in the Shuttle's payload bay for return to Earth. The author describes crew members, daily activities, launch preparation, and payloads during the 11-day mission.

Schuiling, Roelof. "STS-72 Mission Report." *Spaceflight*, April 1996. This article discusses STS-72, Space Shuttle *Endeavour*'s mission to retrieve the Japanese Space Flyer Unit satellite and to deploy and retrieve the Office of Aeronautics and Space Technology Flyer spacecraft. The author describes crew members, daily activities, launch preparation, and payloads during the 10-day mission.

Schuiling, Roelof. "STS-75 Mission Report." *Spaceflight*, June 1996. This article discusses STS-75, Space Shuttle *Columbia*'s mission to deploy and retrieve the joint U.S.-Italian

Tethered Satellite System. The crew had deployed the satellite and had begun gathering scientific data when the tether snapped on flight day three, as the satellite was just short of full deployment by about 12.8 miles. The author describes crew members, daily activities, launch preparation, and payloads during the 15-day mission.

Stine, Deborah D. "U.S. Civilian Policy Priorities: Reflections 50 Years After Sputnik." CRS Report for Congress, Congressional Research Service, Library of Congress, Washington, DC, 2 February 2009. *http://www.fas.org/sgp/crs/space/RL34263.pdf* (accessed 7 March 2012). This report explains how the actions of other nations—the Soviet Union's launch of Sputnik, for example—and the actions of U.S. commercial organizations have influenced U.S. civilian space policy today. The report concludes with a discussion of possible priorities for future U.S. civilian space policy.

U.S. Congress. House of Representatives. Committee on Science. *U.S.-Japanese Cooperation in Human Spaceflight*. 104th Cong., 1st sess., 19 October 1995. *https://web.lexis-nexis.com/ congcomp/* (accessed 11 November 2011 via Proquest Congressional). This hearing provides background on U.S.-Japanese cooperative space programs, including the status of these programs and the outlook for their future. Specifics of Japanese participation in ISS scientific experiments are also included.

Williamson, Ray A. "The US-Europe Technology Gap in Space Transportation: The View from the USA." *Space Policy* 17, no. 1 (February 2001): 27–33. This article reports that competition with Europe's Ariane launcher has influenced the U.S. decision to privatize the expendable launch vehicle production and operation industry. The author also reports that the U.S. government has decided to fly all government payloads on the Space Shuttle and to market Space Shuttle payloads to private industry.

Link to Part 1 (1970–1991), Chapter 13—The Shuttle in International Perspective

CHAPTER 13—MANAGEMENT OF THE SPACE SHUTTLE PROGRAM

Abbey, George, and Neal Lane. *United States Space Policy: Challenges and Opportunities.* Cambridge, MA: American Academy of Arts and Sciences, 2005. *http://carnegie.org/ fileadmin/Media/Publications/PDF/spaceUS.pdf* (accessed 3 April 2012). This publication identifies challenges and opportunities for the U.S. space program, paying particular attention to unintended consequences of current policies. The authors recommend extending the scheduled end date of Space Shuttle missions from 2010 to 2015, so that NASA will not lose the capability of piloted spaceflight.

"At 15, a Safer, Cheaper Shuttle." *Aviation Week & Space Technology*, 8 April 1996. This article examines the Space Shuttle program's status at the fifteenth anniversary of its creation in 1996. Comparing the current space program to the program originally envisioned—as an inexpensive means of access to space—the author discusses how the program has changed over time, explaining how it has failed to meet its early goals.

Blomberg, Richard D. "Report on Shuttle Safety." *Ad Astra*, August 2002. This article summarizes the Space Shuttle safety issues addressed by NASA's Aerospace Safety Advisory Panel (ASAP), along with the panel's findings and recommendations regarding Space Shuttle plans and budgetary requests. The author also makes recommendations regarding ground infrastructure and launch workforce.

Covault, Craig. "Flight of the Phoenix." *Aviation Week & Space Technology*, 11 July 2005. This article discusses the repair and maintenance of Space Shuttle *Discovery* by NASA and contractor personnel. The author reports that the immediate fate of thousands of Shuttle-related aerospace jobs across the United States is riding on the success or failure of *Discovery*'s return to flight and subsequent Shuttle missions.

Covault, Craig. "NASA's Eroding Safety." *Aviation Week & Space Technology*, 12 May 2003. Covault predicts that the *Columbia* Accident Investigation Board's (CAIB's) report will cite serious deficiencies in NASA's overall safety program as a root cause or significant contributing factor to the loss of Space Shuttle *Columbia* and its crew. The author anticipates that the report will question whether NASA's oversight of the spacecraft was effective, once NASA had transferred specific duties of vehicle quality control to the United Space Alliance, as required under the Space Flight Operations Contract.

Covault, Craig. "New Shuttle Concerns Aired." *Aviation Week & Space Technology*, 15 July 1996. The author reports that the White House, NASA, and the aerospace industry are assessing the effect of an independent Aerospace Safety Advisory Panel (ASAP) report that raises serious Space Shuttle safety issues, including the potential for increased risk of a Shuttle accident. The concerns stem from NASA's plan to reduce its costs by shifting additional operational responsibilities to the commercial contractor United Space Alliance. NASA and Thiokol are in the midst of a major investigation of the Space Shuttle's solid rocket motors to determine why hot gas penetrated into new areas of all six-field joints on the two boosters that launched the orbiter *Columbia*.

Covault, Craig. "Shuttle Shakeup Eyed for Cost, Safety Goals." *Aviation Week & Space Technology*, 23 September 2002. This article describes NASA's efforts to reform management and procurement policies and practices for all U.S. piloted space projects. In replacing the Shuttle, NASA hopes to shift from privatized to competitive sourcing. However, NASA has discovered that these efforts must be closely joined to ISS and Space Launch Initiative developments.

Figliola, Patricia Moloney, Carl E. Behrens, and Daniel Morgan. "U.S. Space Programs: Civilian, Military, and Commercial." CRS Issue Brief for Congress, Congressional Research Service, Library of Congress, Washington, DC, 13 June 2006. *http://digital. library.unt.edu/ark:/67531/metacrs10507/?q=IB92011* (accessed 8 March 2012). This report discusses how to manage DOD space programs, avoiding the cost growth and schedule delays that have characterized several recent projects. The authors also describe the appropriate role of the government in facilitating commercial space businesses.

Lawler, Andrew. "NASA Budget Soars as Space Shuttle Lands." *Science* 309, no. 5734 (22 July 2005): 540–541. This article reports the increase of NASA's 2007 budget by approximately US$1 billion.

Lawler, Andrew. "NASA May Cut Shuttle Flights and Reduce Science on Station." *Science* 309, no. 5734 (22 July 2005): 540–541. The article focuses on NASA's plan to curtail Space Shuttle flights and to reduce science research activities on the ISS.

Lawler, Andrew. "Rising Cost of Shuttle and Hubble Could Break NASA Budget." *Science* 305, no. 5692 (24 September 2004): 1882–1883. This article reports that NASA's Associate Administrator of Science, Alphonso V. Diaz, has directed his managers to reduce space and Earth science programs by US$400 million, so that NASA can resume Space Shuttle flights.

Morgan, Daniel, and Carl E. Behrens. "National Aeronautics and Space Administration: Overview, FY 2008 Budget in Brief, and Key Issues for Congress." CRS Report for Congress, Congressional Research Service, Library of Congress, Washington, DC, 14 March 2007. *http://assets.opencrs.com/rpts/RS22625_20070314.pdf* (accessed 8 March 2012). This report discusses the FY 2008 US$17.309 billion budget request for NASA, an increase of 6.5 percent from the FY 2007 appropriation of US$16.247 billion. Other issues addressed include the President's Vision for Space Exploration, development of new vehicles for human spaceflight, plans for the transition to these vehicles after NASA retires the Space Shuttle in 2010, and NASA's efforts to balance its priorities between human exploration and its other activities in science and aeronautics.

Morgan, Daniel, and Carl E. Behrens. "National Aeronautics and Space Administration: Overview, FY 2009 Budget, and Issues for Congress." CRS Report for Congress, Congressional Research Service, Library of Congress, Washington, DC, 26 February 2008. *http://assets.opencrs.com/rpts/RS22818_20080226.pdf* (accessed 8 March 2012). This report discusses NASA's FY 2009 budget request of US$17.614 billion, an increase of 1.8 percent from the FY 2008 appropriation of US$17.309 billion. The report explains the importance of implementing the President's Vision for Space Exploration, including the development of new vehicles for human spaceflight, plans for the transition to these vehicles after NASA retires the Space Shuttle in 2010, and NASA's efforts to balance its priorities between human exploration and its other activities in science and aeronautics.

McCurdy, Howard E. "The Cost of Space Flight." *Space Policy* 10, no. 4 (November 1994): 277–289. The author discusses the costs of the human spaceflight program and NASA's efforts to reduce costs.

Morring, Frank, Jr. "Off the Ground; NASA Included Foreign Rockets, Shuttle-Derived Vehicles as Possible Launchers for Exploration." *Aviation Week & Space Technology*, 28 June 2004. This article explains how contractors could cobble together exploration launch vehicles from Space Shuttle components to save development costs. For instance, NASA calculates that an unpiloted vehicle based on Space Shuttle solid rocket motors, Space Shuttle main engines (SSMEs), and other hardware could lift from 60 to 100 tonnes (66 to 110 tons).

"NASA Loses Head After Budget Boost." *Interavia Business & Technology*, Winter 2004. The article describes congressional budget increases at NASA. Congressional conferees voice concern over future cost overruns on the ISS and how overruns will affect the Space Shuttle.

"Old, Unsafe, and Costly." *Economist*, 30 August 2003. The article comments on proposals to discontinue the Space Shuttle program because of problems in Shuttle design, the costs of the program, and the safety risk of launching these space vehicles into space. In addition, the article discusses the possibility that the Space Shuttle program is having a negative effect on the development of a private space industry. The author believes that the Shuttle program has failed and that NASA should concentrate on developing high-risk technologies with the potential to transform routine space travel for people and equipment.

Pielke, Roger A., Jr. "A Reappraisal of the Space Shuttle Programme." *Space Policy* 9, no. 2 (12 February 2003): 133–157. Congressional and presidential support for the Space Shuttle has been consistently generous despite NASA's inconsistent and flawed justifications for the program. NASA needs to have more rigorous congressional oversight and to develop smaller, quicker, and more independent civil space programs.

Pielke, Roger A., Jr. "Space Shuttle Value Open to Interpretation." *Aviation Week & Space Technology*, 26 July 1993. The article explains Space Shuttle costs from the point of view of a taxpayer, a space policymaker, and a national policymaker.

Schwartz, John. "Report Says Space Program Is Lacking Money and Focus." *New York Times*, 23 June 2005. Two influential experts, George W. S. Abbey, Director of NASA's Johnson Space Center in Houston from 1995 to 2001, and Neal F. Lane, Science Advisor to President William J. Clinton from 1998 to 2001, predict that the President George W. Bush administration's plans for human space exploration are doomed to failure without a major infusion of money and fundamental changes in space policy.

Sietzen, Frank, Jr. "The Future of Space Transportation: Is It Expendable? Reusables? Or the Shuttle? Why Not All Three?" *Ad Astra*, August 2002. This short article discusses the future of the Space Shuttle and Congress's failure to increase NASA's budget, despite rising ISS costs.

Smith, Marcia S. "National Aeronautics and Space Administration: Overview, FY 2004 Budget in Brief, and Issues for Congress." CRS Report for Congress, Congressional Research Service, Library of Congress, Washington, DC, 23 June 2003. *http://assets.opencrs.com/ rpts/RS21430_20030728.pdf* (accessed 12 March 2012). This report focuses on NASA's US$15.5 billion FY 2004 budget request. The author discusses the investigation of the Space Shuttle *Columbia* tragedy on 1 February 2003 and its implications for NASA and for the space program as a whole.

Smith, Marcia S., and Daniel Morgan. "The National Aeronautics and Space Administration: Overview, FY 2005 Budget in Brief, and Key Issues for Congress." CRS Report for Congress, Congressional Research Service, Library of Congress, Washington, DC, 5 October 2004. *http://www.fas.org/spp/civil/crs/RS21744.pdf* (accessed 8 March 2012). This report describes NASA's FY 2005 budget request for US$16.2 billion, a 5.6 percent increase over its FY 2004 appropriation of US$15.4 billion. The report also discusses the new space exploration goals that President George W. Bush announced on 14 January 2004, NASA's plans to return the Space Shuttle to flight status following the Space Shuttle *Columbia* accident, and the health of NASA's aeronautics research program.

Smith, Marcia S., and Daniel Morgan. "The National Aeronautics and Space Administration's FY 2005 Budget Request: Description, Analysis, and Issues for Congress." CRS Report for Congress, Congressional Research Service, Library of Congress, Washington, DC, 12 October 2004. *http://assets.opencrs.com/rpts/RL32676_20041210.pdf* (accessed 8 March 2012). The report describes NASA's FY 2005 budget of US$16.070 billion, a 4.5 percent increase over NASA's FY 2004 appropriation of US$15.378 billion. According to President George W. Bush's Vision for Space Exploration, NASA will focus its activities on returning humans to the Moon by 2020 and someday sending them to Mars and to "worlds beyond."

Smith, Marcia S., and Daniel Morgan. "The National Aeronautics and Space Administration's FY 2006 Budget Request: Description, Analysis, and Issues for Congress." CRS Report for Congress, Congressional Research Service, Library of Congress, Washington, DC, 17 November 2005. *http://assets.opencrs.com/rpts/RL32988_20051117.pdf* (accessed 8 March 2012). This report describes NASA's FY 2006 budget and the congressional debate over NASA's future programs. NASA requested US$16.456 billion, 2.4 percent more than the US$16.070 billion Congress appropriated in FY 2005. NASA Administrator Michael D. Griffin is accelerating development of a crew exploration vehicle.

Smith, Marcia S., Daniel Morgan, and Wendy H. Schacht. "The National Aeronautics and Space Administration's FY 2004 Budget Request: Description, Analysis, and Issues for Congress." CRS Report for Congress, Congressional Research Service, Library of Congress, Washington, DC, 23 September 2003. *http://assets.opencrs.com/rpts/*

RL31821_20030923.pdf (accessed 8 March 2012). This report discusses NASA's budget request of US$15.469 billion for FY 2004, which is approximately 1 percent more than its FY 2003 appropriations level of US$15.339 billion. NASA is making this budget request against the backdrop of the Space Shuttle *Columbia* tragedy, a context that could significantly influence NASA's appropriation.

Trabucco, Peter. "What's Next for NASA After the Space Shuttle?" *Ad Astra*, Fall 2010. This article focuses on the state of the U.S. space program and the future of NASA. As a result of space program budget cuts, employees of NASA's Johnson Space Center could lose their jobs after the last Shuttle flight returns from space.

U.S. Congress. Congressional Budget Office. *A Budgetary Analysis of NASA's New Vision for Space Exploration.* Report, Washington, DC, 2 September 2004. *http://www.cbo.gov/ftpdocs/57xx/doc5772/09-02-NASA.pdf* (accessed 8 March 2012). In this report analyzing NASA's budget request for FY 2005 and NASA's budget projections through 2020, the CBO assesses the implications of NASA's budget plans on the content and schedule of NASA's future activities, including the operation of the Space Shuttle and the United States' participation in the ISS. Funding would enable NASA to develop new vehicles for spaceflight, allowing humans to return to the Moon by 2020.

U.S. Congress. Congressional Budget Office. *Alternatives for Future U.S. Space-Launch Capabilities.* Report, Washington, DC, October 2006. *http://www.cbo.gov/ftpdocs/76xx/doc7635/10-09-SpaceLaunch.pdf* (accessed 8 March 2012). NASA's plan to use piloted spacecraft to return to the Moon by 2020, pursuant to President George W. Bush's 2004 Vision for U.S. Space Exploration, could require the development of the capacity to launch payloads weighing more than 100 tonnes (110 tons). Currently, the payload capacity of launch vehicles does not exceed 25 tonnes (27.6 tons). This report evaluates the feasibility and cost of various alternatives that could significantly increase launch capability for piloted spaceflight beyond low Earth orbit.

U.S. Congress. Congressional Budget Office. *Reinventing NASA.* Report, Washington, DC, 1 March 1994. *http://www.cbo.gov/ftpdocs/48xx/doc4893/doc20.pdf* (accessed 8 March 2012). This report examines NASA's two-pronged strategy to reinvent its program within the confines of a five-year budget plan that is approximately US$25 billion lower than anticipated costs. The CBO also evaluates a set of alternatives that would focus NASA's program more tightly on one or another of its three major traditional objectives—piloted exploration of space, the generation of new scientific knowledge, or the development of space and aeronautical technology.

U.S. Congress. House of Representatives. Committee on Science. *The Future of NASA.* 109th Cong., 1st sess., 28 June 2005. *http://www.gpo.gov/fdsys/pkg/CHRG-109hhrg21949/pdf/CHRG-109hhrg21949.pdf* (accessed 30 March 2012). NASA Administrator Michael D. Griffin is the sole witness at this hearing. His testimony concerns NASA's plans and priorities regarding the ISS, the Space Shuttles' return to flight, the Shuttles' planned retirement in 2010, and the development of crew exploration vehicles.

U.S. Congress. House of Representatives. Committee on Science. *NASA's Fiscal Year 2004 Budget Request.* 108th Cong., 1st sess., 27 February 2003. *http://www.gpo.gov/fdsys/pkg/ CHRG-108hhrg85091/pdf/CHRG-108hhrg85091.pdf* (accessed 30 March 2012). This hearing, which provides an overview of NASA's FY 2004 budget request, covers issues related to NASA's personnel management and its programs, including the status of the ISS and Space Shuttle programs.

U.S. Congress. House of Representatives. Committee on Science. *NASA's Fiscal Year 2006 Budget Proposal.* 109th Cong., 1st sess., 17 February 2005. *http://www.gpo.gov/fdsys/pkg/ CHRG-109hhrg98564/html/CHRG-109hhrg98564.htm* (accessed 30 March 2012). This hearing, which provides an overview of NASA's FY 2006 budget request, covers issues related to NASA programs, including the status of the ISS, the Space Shuttle, the HST, and the development of the crew exploration vehicle.

U.S. Congress. House of Representatives. Committee on Science. *NASA's Fiscal Year 2007 Budget Proposal.* 109th Cong., 2nd sess., 16 February 2006. *http://www.gpo.gov/fdsys/pkg/ CHRG-109hhrg25937/pdf/CHRG-109hhrg25937.pdf* (accessed 12 March 2012). This hearing presents an overview of NASA's FY 2007 budget request, covering issues related to NASA programs, including the status of the ISS, the Space Shuttle, and crew exploration vehicle programs.

U.S. Congress. House of Representatives. Committee on Science. *Status of NASA's Programs.* 109th Cong., 1st sess., 3 November 2005. *http://www.gpo.gov/fdsys/pkg/CHRG-109hhrg24151/pdf/CHRG-109hhrg24151.pdf* (accessed 12 March 2012). This hearing provides an overview of NASA's approach to implementing the new Vision for Space Exploration and reviews issues related to NASA programs, including the status of the ISS, the Space Shuttle, and crew exploration vehicle programs.

U.S. Congress. House of Representatives. Committee on Science. Subcommittee on Space and Aeronautics. *Determinations and Findings for the Space Shuttle Program.* 104th Cong., 1st sess., 30 November 1995. *https://web.lexis-nexis.com/congcomp/* (accessed 11 November 2011 via Proquest Congressional). This hearing examines NASA's decisions to suspend competitive bidding for Space Shuttle contracts and to negotiate a new contract for consolidating Space Shuttle operations under a single prime contractor, the United Space Alliance, a joint venture of Lockheed Martin and Rockwell International.

U.S. Congress. House of Representatives. Committee on Science. Subcommittee on Space and Aeronautics. *Fiscal Year 1996 NASA Authorization.* 104th Cong., 1st sess., 13 February and 16 March 1995. *https://web.lexis-nexis.com/congcomp/* (accessed 11 November 2011 via Proquest Congressional). These hearings review cutbacks to NASA's budget and the effects of these decreases on program priorities. The hearings also provide an overview of Space Shuttle and ISS operations, including safety programs.

U.S. Congress. House of Representatives. Committee on Science. Subcommittee on Space and Aeronautics. *Fiscal Year 1997 NASA Authorization.* 104th Cong., 2nd sess., 17 April 1996. *https://web.lexis-nexis.com/congcomp/* (accessed 11 November 2011 via Proquest

Congressional). This hearing provides an overview of the FY 1997 budget for the Space Shuttle program and reviews the status of the space station and Shuttle programs. Testimony discusses negotiations between NASA and United Space Alliance regarding a contract for spaceflight operations for the Space Shuttle. The hearing also reviews aspects of NASA's safety programs involving the Shuttle.

U.S. Congress. House of Representatives. Committee on Science. Subcommittee on Space and Aeronautics. *Fiscal Year 2001 NASA Authorization: NASA Posture, Parts I–VI.* 106th Cong., 2nd sess., 16 February, 16 and 22 March, 11 April, 10 May, and 13 September 2000. *https://web.lexis-nexis.com/congcomp/* (accessed 11 November 2011 via Proquest Congressional). This hearing presents NASA's FY 2001 budget request for human spaceflight, focusing on the ISS and Space Shuttle programs. Testimony reviews the findings of OPM's Office of the Inspector General, Office of Audits, which examined NASA human spaceflight programs, projects, and activities.

U.S. Congress. House of Representatives. Committee on Science. Subcommittee on Space and Aeronautics. *NASA's Fiscal Year 1999 Budget Request, Parts I–IV.* 105th Cong., 2nd sess., 5–25 February and 19 March 1998. *https://web.lexis-nexis.com/congcomp/* (accessed 11 November 2011 via Proquest Congressional). This hearing presents NASA's FY 1999 budget request for human exploration and the development of the space enterprise, focusing on the priorities for and the costs of the ISS and Space Shuttle programs.

U.S. Congress. House of Representatives. Committee on Science. Subcommittee on Space and Aeronautics. *Space Shuttle Program in Transition: Keeping Safety Paramount, Parts I and II.* 104th Cong., 1st sess., 27 September and 9 November 1995. *https://web.lexis-nexis.com/congcomp/* (accessed 11 November 2011 via Proquest Congressional). These hearings review NASA's policies for ensuring the flight safety of the Space Shuttle program in light of NASA's planned transfer of program operations to a single prime contractor.

U.S. Congress. House of Representatives. Committee on Science. Subcommittee on Space and Aeronautics. *Space Shuttle Safety.* 105th Cong., 1st sess., 1 October 1997. *https://web.lexis-nexis.com/congcomp/* (accessed 11 November 2011 via Proquest Congressional). This hearing provides a justification for NASA's transfers of budgeted funds from the Space Shuttle program to its other programs; reviews the safety of the Shuttle program, including the effect of the transfer of funds on Shuttle safety; and presents views on Shuttle costs and improvements.

U.S. Congress. House of Representatives. Committee on Science and Technology. *NASA's Fiscal Year 2008 Budget Request.* 110th Cong., 1st sess., 15 March 2007. *http://frwebgate.access.gpo.gov/cgi-bin/getdoc.cgi?dbname=110_house_hearings&docid=f:33803.pdf* (accessed 13 March 2012). This hearing, an overview of NASA's FY 2008 budget request, focuses on issues related to NASA programs, including the status of the ISS, the Space Shuttle, and crew exploration vehicle programs. The hearing also discusses concerns about potential funding shortfalls in NASA's FY 2008 budget request and the effect of these shortfalls on NASA programs.

U.S. Congress. House of Representatives. Committee on Science and Technology. *NASA's Fiscal Year 2009 Budget Request*. 110th Cong., 2nd sess., 13 February 2008. *http://www.gpo.gov/fdsys/pkg/CHRG-110hhrg40598/html/CHRG-110hhrg40598.htm* (accessed 8 March 2012). This hearing, an overview of NASA's FY 2009 budget request, covers issues related to NASA programs, including the status of the ISS, the Space Shuttle, and exploration and research programs. The request includes US$2.98 billion to operate and maintain NASA's three Space Shuttles.

U.S. Congress. House of Representatives. Committee on Science and Technology. *NASA's Fiscal Year 2010 Budget Request*. 111th Cong., 1st sess., 19 May 2009. *http://www.gpo.gov/fdsys/pkg/CHRG-111hhrg49551/pdf/CHRG-111hhrg49551.pdf* (accessed 13 March 2012). This hearing, an overview of NASA's FY 2010 budget request, covers issues related to NASA programs, including the status of the ISS, the Space Shuttle, and exploration and research programs.

U.S. Congress. House of Representatives. Committee on Science and Technology. *NASA's Fiscal Year 2011 Budget Request and Issues*. 111th Cong., 2nd sess., 25 February 2010. *http://frwebgate.access.gpo.gov/cgi-bin/getdoc.cgi?dbname=111_house_hearings& docid=f:55837.pdf* (accessed 8 March 2012). This hearing examines NASA's FY 2011 budget request, covering issues related to NASA programs, including the status of the ISS, the Space Shuttle, and exploration and research programs.

U.S. Congress. House of Representatives. Committee on Science, Space, and Technology. Subcommittee on Space. *1992 NASA Authorization, Volume II*. 102nd Cong., 1st sess., 7 February–11 April 1991. *https://web.lexis-nexis.com/congcomp/* (accessed 11 November 2011 via Proquest Congressional). These hearings include a presentation of NASA's FY 1992 budget request for space transportation system programs, including the Space Shuttle and launch systems, and an explanation of the restructured ISS program.

U.S. Congress. House of Representatives. Committee on Science, Space, and Technology. Subcommittee on Space. *1995 NASA Authorization*. 103rd Cong., 2nd sess., 23 February, 23 March, and 14 April 1994. *https://web.lexis-nexis.com/congcomp/* (accessed 11 November 2011 via Proquest Congressional). These hearings provide an explanation of NASA's FY 1995 budget request, focusing on efforts to maintain key program priorities in spite of budget cutbacks. The testimony includes an explanation of the budget request for human spaceflight programs, including the ISS and Space Shuttle.

U.S. Congress. House of Representatives. Committee on Science, Space, and Technology. Subcommittee on Space. *Contract Management Issues: Cost Overruns on NASA's Shuttle Toilet*. 103rd Cong., 1st sess., 23 February 1993. *https://web.lexis-nexis.com/congcomp/* (accessed 11 November 2011 via Proquest Congressional). This hearing reviews the complexities and importance of developing a human waste collection system (WCS) for use on the Space Shuttle *Endeavour*, reviews NASA cost management operations regarding the Rockwell International Corporation contract, and provides an explanation for the wide discrepancy between the originally estimated cost and the actual cost of developing the WCS.

U.S. Congress. Senate. Committee on Commerce, Science, and Transportation. Subcommittee on Science. *NASA's Fiscal Year 1992 Budget Overview.* 102nd Cong., 1st sess., 19 April 1991. *https://web.lexis-nexis.com/congcomp/* (accessed 11 November 2011 via Proquest Congressional). This hearing, an overview of the FY 1992 budget request for NASA, including funding for its space transportation and exploration programs, provides a perspective on NASA's priorities and concerns regarding the Space Shuttle and ISS programs.

U.S. Congress. Senate. Committee on Commerce, Science, and Transportation. Subcommittee on Science and Space. *Human Spaceflight: The Space Shuttle and Beyond.* 109th Cong., 1st sess., 18 May 2005. *http://www.gpo.gov/fdsys/pkg/CHRG-109shrg25323/pdf/CHRG-109shrg25323.pdf* (accessed 30 March 2012). This hearing reviews Space Shuttle operations and future human spaceflight issues, including concerns for the safety and the value of the human spaceflight program. It also covers the planned retirement of the Shuttle fleet, NASA's transition to a new launch system, and the effect of these changes on NASA's workforce.

U.S. Congress. Senate. Committee on Commerce, Science, and Transportation. Subcommittee on Science, Technology, and Space. *NASA Management Problems.* 106th Cong., 2nd sess., 22 March 2000. *http://www.gpo.gov/fdsys/pkg/CHRG-106shrg78634/pdf/CHRG-106shrg78634.pdf* (accessed 30 March 2012). In this hearing, which examines NASA's management problems and reform initiatives, NASA Administrator Daniel S. Goldin and others discuss the results of the Space Shuttle Independent Assessment Team's review of Space Shuttle subsystems and maintenance practices. The hearing also reviews the findings of the GAO's report assessing issues associated with the Space Shuttle program's civil service workforce.

U.S. Congress. Senate. Committee on Commerce, Science, and Transportation. Subcommittee on Science, Technology, and Space. *NASA Space Shuttle and the Reusable Launch Vehicle Programs.* 104th Cong., 1st sess., 16 May 1995. *https://web.lexis-nexis.com/congcomp/* (accessed 11 November 2011 via Proquest Congressional). This hearing, an overview of the organization and policy of NASA's reusable space launch vehicle programs, reviews possibilities for restructuring and streamlining Space Shuttle program operations, such as placing greater reliance on contractors.

U.S. Congress. Senate. Committee on Commerce, Science, and Transportation. Subcommittee on Science, Technology, and Space. *Shuttle Safety.* 107th Cong., 1st sess., 6 September 2001. *http://www.gpo.gov/fdsys/pkg/CHRG-107shrg82708/pdf/CHRG-107shrg82708.pdf* (accessed 30 March 2012). This hearing reviews NASA's efforts to implement Space Shuttle program safety and performance upgrades, specifically focusing on concerns about the effect of proposed budget cuts on NASA's ability to implement Space Shuttle safety upgrades and infrastructure improvements. The hearing also covers NASA's efforts to address workforce issues relating to the Space Shuttle program.

U.S. Congress. Senate. Committee on Commerce, Science, and Transportation. Subcommittee on Science, Technology, and Space. *Space Station and Space Shuttle Programs*. 104th Cong., 2nd sess., 24 July 1996. *https://web.lexis-nexis.com/congcomp/* (accessed 11 November 2011 via Proquest Congressional). This hearing reviews the status of NASA's Space Shuttle and ISS programs, including issues involved in cost control and schedule management of the space station program.

U.S. General Accounting Office. "Federal R&D Laboratories." Report no. GAO/RCED/NSIAD-96-78R, Washington, DC, October 29, 1996. *http://archive.gao.gov/papr2pdf/156329.pdf* (accessed 8 March 2012). This report provides FY 2005 operating budgets for Shuttle activities at NASA's Ames Research Center, Goddard Space Flight Center, Johnson Space Center, Kennedy Space Center, Langley Research Center, Lewis Research Center, Marshall Space Flight Center, and Stennis Space Center.

U.S. General Accounting Office. "International Space Station and Shuttle Support Cost Limits." Briefing to Staff of the Senate Committee on Commerce, Science, and Transportation and House Committee on Science. Report no. GAO-04-648R, Washington, DC, 2 April 2004. *http://www.gao.gov/new.items/d04648r.pdf* (accessed 8 March 2012). This report analyzes issues relating to NASA's inadequate reporting of amounts obligated toward ISS and Space Shuttle spending, as required under the NASA Authorization Act of 2000, which limits NASA's spending on these programs.

U.S. General Accounting Office. "Large Programs May Consume Increasing Share of Limited Future Budgets." Report no. GAO/NSIAD-92-278, Washington, DC, 9 April 1992. *http://archive.gao.gov/d35t11/147504.pdf* (accessed 8 March 2012). This report outlines Space Shuttle costs, including production and operation costs. The GAO recommends that, because the Space Shuttle program accounts for such a large portion of NASA's budget, Congress should consider directing NASA's Administrator to incorporate five-year program estimates and life-cycle costs into the funding section of NASA's biannual report on the project's status.

U.S. General Accounting Office. "Major Management Challenges and Program Risks: National Aeronautics and Space Administration." Report no. GAO-01-258, Washington, DC, January 2001. *http://www.gao.gov/pas/2001/d01258.pdf* (accessed 8 March 2012). In this report, the GAO indicates that, since 1995, the Space Shuttle workforce has decreased by more than one-third. Many key areas of the program have insufficient qualified staff, and the workforce shows signs of overwork and fatigue. Moreover, the skill mix and demographics of the Shuttle workforce jeopardize NASA's ability to increase the Shuttle flight rate in support of the ISS's assembly and to transfer leadership roles to the next generation.

U.S. General Accounting Office. "Major Management Challenges and Program Risks: National Aeronautics and Space Administration." Report no. GAO-03-114, Washington, DC, 1 January 2003. *http://www.gao.gov/new.items/d03114.pdf* (accessed 8 March 2012). This report focuses on management, oversight, and workforce issues facing NASA. According to the GAO, NASA needs to transform its long-standing business practices, enabling it to strengthen its strategic human capital management, reduce space launch costs, and improve contract management.

U.S. General Accounting Office. "NASA Budgets: Gap Between Funding Requirements and Projected Budgets Has Been Reopened." Report no. GAO/NSIAD-95-155BR, Washington, DC, May 1995. *http://www.gao.gov/archive/1995/ns95155b.pdf* (accessed 8 March 2012). This report focuses on the discrepancy between NASA's program plans and its likely budget. The projected gap for fiscal years 1996 through 2000 is US$5.3 billion. When the GAO reported such a gap in 1992, NASA changed or deleted some of its major programs to help eliminate the discrepancy. However, those changes resulted in increased risks in several of its largest programs.

U.S. General Accounting Office. "NASA: Compliance with Cost Limits Cannot Be Verified." Report no. GAO-02-504R, Washington, DC, 10 April 2002. *http://www.gao.gov/new. items/d02504r.pdf* (accessed 8 March 2012). This report examines the congressionally imposed limits on NASA's FY 2002 spending on ISS development and Space Shuttle flights. NASA has acknowledged that it lacks a modern, integrated financial management system capable of providing the detailed data needed to support amounts obligated against the limits.

U.S. General Accounting Office. "NASA: International Space Station and Shuttle Support Cost Limits." Report no. GAO-01-1000R, Washington, DC, 31 August 2001. *http://www.gao. gov/new.items/d011000r.pdf* (accessed 8 March 2012). This report analyzes NASA's FY 2000 budget authorization. Because NASA was unable to furnish financial data in support of the actual cost to complete space station elements and subsystems, the GAO could not determine whether NASA's costs were reasonable.

U.S. General Accounting Office. "NASA Issues." Report no. GAO/OCG-93-27TR, Washington, DC, 12 January 1992. *http://archive.gao.gov/d36t11/148271.pdf* (accessed 8 March 2012). This report claims that NASA's strategic and program plans feature unrealistic funding levels and recommends that NASA improve operations management and oversight of its Space Shuttle program to enhance its efficiency and effectiveness.

U.S. General Accounting Office. "Space Shuttle: Further Improvements Needed in NASA's Modernization Efforts." Report no. GAO-04-203, Washington, DC, 15 January 2004. *http://www.gao.gov/new.items/d04203.pdf* (accessed 8 March 2012). In this report, the GAO recommends that the NASA Administrator fully define the requirements for all elements of the Integrated Space Transportation Plan. Specifically, the report recommends that NASA and the ISS partners determine the ultimate life and mission of the space station, so that NASA will have a sound basis for fully defining Space Shuttle requirements.

U.S. General Accounting Office. "Space Shuttle: Human Capital and Safety Upgrade Challenges Require Continued Attention." Report no. GAO/NSIAD/GGD-00-186, Washington, DC, 15 August 2000. *http://www.gao.gov/archive/2000/n200186.pdf* (accessed 8 March 2012). This report discusses several internal NASA studies showing that downsizing has harmed the Space Shuttle program's workforce, posing significant risks to Shuttle flight safety. Many key areas of the program have insufficient qualified staff, and the workforce shows signs of overwork and fatigue. Furthermore, the age of NASA's workforce—more than twice as many workers over the age of 60 as under the age of 30—jeopardizes NASA's ability to transfer leadership to the next generation.

U.S. General Accounting Office. "Space Shuttle: Incomplete Data and Funding Approach Cost Risk for Upgrade Program." Report no. GAO/NSIAD-94-23, Washington, DC, 26 May 1994. *http://archive.gao.gov/t2pbat3/151985.pdf* (accessed 8 March 2012). This report indicates that NASA's budget cannot simultaneously support development of a new launch system and construction of the ISS. The report recommends that NASA's Administrator direct Space Shuttle program managers to estimate the life-cycle costs of proposed Shuttle upgrades.

U.S. General Accounting Office. "Space Shuttle: NASA Must Reduce Costs Further To Operate Within Future Projected Funds." Report no. GAO/NSIAD-95-118, Washington, DC, June 1995. *http://archive.gao.gov/t2pbat1/154853.pdf* (accessed 8 March 2012). This report assesses Space Shuttle program cost reductions, past and future, and the effects of these reductions on safety. So far, NASA has reduced its cumulative funding for Shuttle operations by 22 percent, from FY 1992 to FY 1995, and its actual operating costs by 8.5 percent, between FY 1992 and 1994. The GAO report found that, although additional funding reductions are necessary to achieve NASA's future budget projections, NASA may not be able to reduce its costs without affecting safety.

U.S. General Accounting Office. "Space Shuttle: NASA's Plans for Repairing or Replacing a Damaged or Destroyed Orbiter." Report no. GAO/NSIAD-94-197, Washington, DC, 21 July 1994. *http://archive.gao.gov/t2pbat2/152401.pdf* (accessed 8 March 2012). In this report, the GAO discusses NASA's lack of a contingency plan in the event that a Space Shuttle sustains damage. NASA has no capability to replace or repair a seriously damaged orbiter. NASA is terminating its structural spares program and intends to finish those parts that are 85 percent or more complete and to place them in storage. NASA does not believe this decision will significantly increase the risk to the Space Shuttle program.

U.S. General Accounting Office. "Space Shuttle: Upgrade Activities and Carryover Balances." Report no. GAO/T-NSIAD-98-21, Washington, DC, 1 October 1997. *http://www.gao.gov/archive/1998/ns98021t.pdf* (accessed 8 March 2012). In this report, the GAO analyzes NASA's plan to take US$190 million from the Space Shuttle program to help offset additional costs of the ISS. The GAO notes that the transfer of the funds to the ISS program does not adversely affect current or near-term Shuttle upgrade projects.

U.S. General Accounting Office. "Space Station: Cost Control Difficulties Continue." Report no. GAO/NSIAD-96-135, Washington, DC, July 1996. *http://www.gao.gov/archive/1996/ns96135.pdf* (accessed 8 March 2012). This report describes cost overruns for the ISS. As of April 1996, the ISS was nearly US$90 million over cost. The GAO warns that, if available resources prove inadequate, NASA will either have to exceed its annual funding limitation on the ISS or defer or rephase other activities, potentially delaying the space station's construction schedule and increasing its overall cost.

U.S. General Accounting Office. "Space Transportation: The Content and Uses of Shuttle Cost Estimates." Report no. GAO/NSIAD-93-115, Washington, DC, 28 January 1993. *http://archive.gao.gov/d37t11/148584.pdf* (accessed 8 March 2012). This report describes costs associated with the operation of the Space Shuttle. The report notes that NASA's reported average cost per flight does not include some of the program's major costs, such as Shuttle development and future Shuttle upgrades.

U.S. Government Accountability Office. "International Space Station and Shuttle Support Cost Limits." Briefing to the Senate Committee on Commerce, Science, and Transportation and the House Committee on Science, Report no. GAO-05-492R, Washington, DC, 8 April 2005. *http://www.gao.gov/new.items/d05492r.pdf* (accessed 8 March 2012). This report examines whether NASA is fulfilling it accounting requirements according to the NASA Authorization Act of 2000. The Act requires that the GAO verify NASA's accounting for amounts obligated against established limits for the space station and related Space Shuttle support.

U.S. Government Accountability Office. "NASA: Long-Term Commitment to and Investment in Space Exploration Program Requires More Knowledge." Report no. GAO-06-817R, Washington, DC, 17 July 2006. *http://www.gao.gov/new.items/d06817r.pdf* (accessed 8 March 2012). This report reviews NASA's plans to spend nearly US$230 billion over the next two decades to implement the President's Vision for Space Exploration. NASA continues to refine cost estimates for its exploration architecture but is unable to provide a firm estimate of the costs of implementation, mainly because the program is in its early stages.

Link to Part 1 (1970–1991), Chapter 14—Management and Funding of the Shuttle Program

CHAPTER 14—JUVENILE LITERATURE ABOUT THE SPACE SHUTTLE

Adamson, Heather. *The Challenger Explosion*. Mankato, MN: Capstone Press, 2006. Ages 9–12. This graphic novel tells the story of Christa McAuliffe and the six other NASA astronauts who lost their lives in the Space Shuttle *Challenger* disaster on 28 January 1986.

Amato, William. *The Space Shuttle (High-Tech Vehicles)*. New York: PowerKids Press, 2002. Ages 4–8. This book for school-age children shows how technologically advanced Space Shuttles are made and used, providing school an inside view of these high-powered, high-tech vehicles.

Baker, David, and Heather Kissock. *The Shuttle*. New York: Weigl, 2009. Ages 4–8. This book, with its simple written text and a mixture of small color photos and artistic renditions of the Space Shuttle, describes the technology used to explore the universe.

Bergin, Mark. *Space Shuttle*. New York: Franklin Watts, 1999. This book, recommended for elementary and junior high school readers, follows the development of the Space Shuttle program from the early years to the present day, covering construction of the Shuttle and missions in space, including takeoff, reentry, and landing.

Blast Off! A Space Track Adventure. Stamford, CT: Innovative Kids, 1999. This book features a windup toy Space Shuttle with its own built-in storage compartment, punch-out cardboard accessories, built-in tracks on every page, and a special foldout with almost two additional feet of track. The reader "rides" the toy Space Shuttle across space on track pages while reading about the universe.

Branley, Franklyn Mansfield. *Floating in Space*. New York: HarperCollins, 1998. Ages 4–8. Grades K–2. The book, filled with kid-friendly diagrams and illustrations, describes life-support systems and weightlessness aboard the Space Shuttle, including descriptions of what astronauts eat, how they move, and what kinds of tasks they do in space.

Bredeson, Carmen. *Astronauts*. New York: Children's Press, 2003. This book introduces young readers to the work of astronauts, including special jobs they might have on the Space Shuttle and experiments they might perform in space.

Bredeson, Carmen. *Getting Ready for Space*. New York: Children's Press, 2003. This book provides a simple overview of the training that astronauts go through for missions on the Space Shuttle.

Bredeson, Carmen. *John Glenn Returns to Orbit: Life on the Space Shuttle*. Berkeley Heights, NJ: Enslow Publishers, 2000. Ages 9 and older. This book describes the activities aboard Space Shuttle *Discovery* during its historic flight in 1998, when astronaut John H. Glenn Jr., the first American to orbit Earth, returned to space at the age of 77.

Bredeson, Carmen. *Liftoff!* New York: Children's Press, 2003. Ages 4–8. Grades K–2. This book shows the Shuttle being readied for flight and brought to the launchpad on the crawler.

The book features exciting text and photos and describes the difficulties of eating, drinking, and walking in space.

Bredeson, Carmen. *Living on a Space Shuttle*. New York: Children's Press, 2003. Grades K–2. This book provides a simple description of how astronauts aboard a Space Shuttle perform everyday activities such as eating, drinking, and sleeping.

Bredeson, Carmen. *Shannon Lucid Space Ambassador*. Brookfield, CT: Millbrook Press, 1998. *http://www.barnesandnoble.com/s/Shannon-Lucid-Space-Ambassador-?keyword= Shannon+Lucid+Space+Ambassador.&store=ebook* (accessed 8 March 2012). This electronic book chronicles the life of astronaut Shannon W. Lucid, from her childhood in Oklahoma, through her various Space Shuttle missions, to her six months aboard the *Mir* space station.

Bredeson, Carmen. *The Challenger Disaster: Tragic Space Flight*. Berkeley Heights, NJ: Enslow Publishers, 1999. Ages 9–12. This book describes the events surrounding the explosion of Space Shuttle *Challenger* in 1986 and the investigation of the disaster, as well as the stories of the seven astronauts who died on the *Challenger*.

Campbell, Peter A. *Launch Day*. Brookfield, CT: Millbrook Press, 1995. Grades 3–6. This book describes for young readers the preparation, launch, and return of a Space Shuttle *Atlantis* flight, describing the Vehicle Assembly Building and the Tractor-Transporter Vehicle, as the rocket is transported to the launchpad. The author includes many statistics to convey the complexity of a Shuttle liftoff, emphasizing the technological aspects of the flight, rather than life on board the Shuttle or the value of the mission.

Caper, William. *The Challenger Space Shuttle Explosion*. New York: Bearport, 2007. This book examines the events surrounding the explosion of Space Shuttle *Challenger* and discusses the significant changes to the Space Shuttle program that resulted from the loss of *Challenger*.

Cole, Michael D. *The Columbia Space Shuttle Disaster: From First Liftoff to Tragic Final Flight*. Rev. ed. Berkeley Heights, NJ: Enslow Publishers, 2003. This book contains technical details about the Space Shuttle, as well as personal information about the astronauts aboard *Columbia* on its final mission, STS-107. Photographs introduce young readers to the crew and help them understand the events that led to the tragic loss of the spacecraft and its crew.

Cole, Michael D. *Columbia: First Flight of the Space Shuttle*. Springfield, NJ: Enslow Publishers, 1995. Grades 4–6. This book chronicles the first exciting flight of Space Shuttle *Columbia* in August 1981. The author also recounts some of the highlights of the United States' piloted spaceflight program.

Dahl, Michael, Alderman, Derrick, and Shea, Denise. *On the Launch Pad: A Counting Book About Rockets*. Minneapolis, MN: Picture Window Books, 2004. Ages 3 and older. This book counts down from twelve to one as a Space Shuttle awaits liftoff. Readers are invited to find hidden numbers on an illustrated activity page.

Davis, Amanda. *Spaceships*. New York: PowerKids Press, 1997. Ages 9–12. This book discusses various spacecraft, including rockets, satellites, and Space Shuttles, and the jobs that they perform.

Fahey, Kathleen. *Challenger and Columbia*. Milwaukee, WI: Gareth Stevens Publishing, 2005. Ages 9 and older. The book describes the events leading up to the loss of Space Shuttles *Challenger* and *Columbia*. Based on detailed descriptions, expert testimony, and firsthand accounts, the author takes an in-depth look at the *Challenger* and *Columbia* tragedies and discusses the lessons learned from them.

Feldman, Heather. *Columbia: The First Space Shuttle*. New York: PowerKids Press, 2003. Ages 9 and older. This book provides an excellent overview of the Space Shuttle program, from *Columbia*'s first mission to the aftermath of the *Challenger* tragedy, explaining how Space Shuttles work and describing the problems scientists encountered in developing the Shuttle.

Follett Software Company and Amazing Media. *Space Shuttle*. Cupertino, CA: KidSoft, 1996. DVD. This DVD program contains a multimedia introduction to the Space Shuttle program, including descriptions of space vehicles, equipment, and crews; orientation and training at NASA's Johnson Space Center; and details of living and working in space on 53 NASA missions.

Gardner, Charlie. *See How They Go: Spaceship*. London: Dorling Kindersley, 2009. This book introduces kids to vehicles that travel into space, from Moon rockets to Space Shuttles. The book contains dynamic images and simple text, as well as a sheet of stickers of featured vehicles.

Gold, Susan Dudley. *To Space and Back: The Story of the Shuttle*. New York: Crestwood House, 1992. Ages 9–12. The author examines the history, uses, and accomplishments of the Space Shuttle program.

Graham, Ian. *Space Travel*. New York: Dorling Kindersley, 2004. This book examines ways humans explore and live in space. The author includes sections on Space Shuttles, space stations, living in space, spacewalks, spacesuits, training for space, and science in space.

Gross, Miriam J. *All About Space Shuttles*. New York: PowerKids Press, 2009. Ages 4–8. This book explains the history and function of Space Shuttles and how they have helped us reach outer space. Chapters cover Shuttle parts, liftoff, returning to Earth, and Space Shuttle uses.

Hansen, Rosanna. *Liftoff! A Space Adventure*. Pleasantville, NY: Reader's Digest Children's Books, 2000. Ages 7–8. This book invites young readers to imagine blasting off into space aboard a Space Shuttle on a flight to help build a space station. The book describes experiencing G-forces when lifting off into space, how it feels to float in microgravity while orbiting the Earth in a Shuttle, suiting up and going outside of the Shuttle to help build the space station, moving around with the help of a jetpack, eating a meal in space, and napping in a sleeping bag strapped to the wall of the crew quarters to prevent floating around.

Holden, Henry M. *The Tragedy of the Space Shuttle Challenger*. Berkeley Heights, NJ: MyReportLinks.com Books, 2004. Ages 9–12. This book describes events surrounding the explosion of Space Shuttle *Challenger* in 1986 and discusses the investigation of this disaster. The author tells the stories of the seven astronauts who died. The book also includes Internet links to related Web sites, source documents, and photographs.

Jackson, Francine. *Space Shuttle*. Outer Space Series. Danbury, CT: Grolier Educational, 1998. This book, written for a juvenile audience, describes the development of the Space Shuttle, its basic structure, and the daily life of astronauts aboard the Shuttle.

Kallen, Stuart A. *Space Shuttles*. Edina, MN: Abdo and Daughters, 1996. Ages 4–8. The author describes the Space Shuttles' physical characteristics and the lives and experiences of crew members while on board a Shuttle, as well as recounting the history of NASA's Space Shuttle program.

Kerrod, Robin. *Space Shuttles. The History of Space Exploration*. Milwaukee, WI: World Almanac Library, 2004. Ages 9–12. This book, which includes many full-page color photographs, describes three Space Shuttle missions that occurred in 1984.

Kettelkamp, Larry. *Living in Space*. New York: Morrow Junior Books, 1993. Ages 9–12. This book provides a brief history of U.S. space exploration before explaining how astronauts currently live and work in space and describing plans for piloted and unpiloted space exploration in the near future. Discussing spaceflights, Space Shuttles, and future plans for space stations, the author considers what humans have achieved toward living in space, as well as discussing hopes for the future.

Kortenkamp, Steve. *Space Shuttles*. Mankato, MN: Capstone Press, 2008. Ages 4–8. This book describes the history and uses of NASA's Space Shuttles, including both the practical uses and the drawbacks of reusable Space Shuttles. The author explains how the Shuttle flies and how it launches satellites. In addition, he describes how astronauts travel into space to make repairs on missions such as that to service the HST. The book also includes information about the two Shuttle disasters.

Langille, Jacqueline, and Bobbie Kalman. *The Space Shuttle*. New York: Crabtree Publishing, 1998. Ages 7–9. This book describes the construction and operation of NASA's Space Shuttle and the Shuttle's role in the future of humans in space, covering such topics as preparation for liftoff, living and working in space, and EVAs (spacewalks).

Lassieur, Allison. *The Space Shuttle*. New York: Children's Press, 2000. Ages 9–12. This book describes the beginnings of the Space Shuttle program and the role of the Shuttle in building a space station. The author covers such topics as the layout of a Shuttle, a day's activities in space, and the *Challenger* disaster. The book includes a glossary and lists relevant books, organizations, and Internet sites.

Lieurance, Suzanne. *The Space Shuttle Challenger Disaster in American History*. Berkeley Heights, NJ: Enslow Publishers, 2001. Ages 9–12. Grades 5–7. After recounting the events of the morning that Space Shuttle *Challenger* exploded, the author places the *Challenger*'s mission in historical context. She summarizes the U.S. space program up to that time, profiles each of the astronauts assigned to the *Challenger* crew, and discusses the selection and training of civilian teacher Christa McAuliffe for the mission.

Littlejohn, Randy. *Life in Outer Space*. New York: Rosen Publishing Group, 2004. Ages 9–12. Grades 5–8. This book describes conditions for astronauts aboard Space Shuttles, space stations, and space capsules, explaining how humans survive in these extreme environments, and discusses possible future developments in space travel.

Maynard, Christopher. *The Space Shuttle*. New York: Kingfisher Books, 1994. Ages 9–12. This fold-out book, which tells about a Space Shuttle mission to rescue a satellite, has flaps that open to reveal background information related to the story.

McNeese, Tim. *The Challenger Disaster*. New York: Children's Press, 2003. This book documenting the *Challenger* disaster chronicles the failures on the part of NASA engineers to prevent the tragic accident and provides information about the mission and its crew members.

Mullane, R. Mike. *Liftoff! An Astronaut's Dream*. Parsippany, NJ: Silver Burdett, 1995. The author, a former astronaut, describes his experiences in space and shares his ideas about the future of spaceflight.

Murray, Peter. *The Space Shuttle*. Mankato, MN: Child's World, 1993. Ages 7–9. Grades 2–6. This book discusses the development and uses of reusable spacecraft, as well as describing the typical flight of a Space Shuttle. The book is also available in Spanish under the title *La Lanzadera Espacial*.

NASA Aerospace Education Services Project. *Space Shuttle: Activities for Primary and Intermediate Students*. Cleveland, Ohio: Aerospace Education Services Project, Lewis Research Center, NASA, 1992. This book provides activities for primary and intermediate students. It was developed to help school teachers describe the Space Shuttle to students and explain how it operates.

Oxlade, Chris. *Space Shuttle*. North Mankato, MN: Thameside Press, 2002. For infants and children of preschool age. This book describes how the different parts of a Space Shuttle fit together, how astronauts live and work in space, what the inside of a Space Shuttle looks like and how it works, and how the Shuttle's engines and booster rockets launch the Shuttle into space.

Paul, Anthony, Belinda Grode Tatum, Lucious G. Tatum, Michael S. Emerson, and Pamela Geagan. *All About the Military/All About the Space Shuttle*. DVD. New York: GT Media, 2006. This DVD program intended for juveniles invites the viewer to imagine he or she is a new astronaut studying the history of space travel, learning about how and why NASA built the Space Shuttle, and exploring the universe. The viewer has the virtual experience of training to work, live, eat, and sleep in space, floating weightlessly in space, exploring all the decks of the Space Shuttle, and sitting in the seat of the Commander of the Shuttle on a mission to repair and return a satellite to Earth.

Rees, Peter. *Secrets of the Space Shuttle*. New York: Children's Press, 2008. Ages 9–11. This book describes the space race, the first trip to the Moon, how Space Shuttles work, space stations, weightlessness, and what life is like for astronauts.

Richardson, Adele. *Space Shuttle*. Mankato, MN: Smart Apple Media, 2000. Ages 4–8. This book examines the development of Space Shuttles, their components, and how humans have used them to explore outer space.

Rouss, Sylvia A. *Reach for the Stars: A Little Torah's Journey*. New York: Devora Publishing, 2004. Grades 3–5. This is a true story of the miniature Torah that Israeli astronaut Ilan Ramon carried with him on Space Shuttle *Columbia*'s mission, STS-107. This unique Torah survived the Holocaust, along with its guardian, Joachim Joseph. Years later, Ilan Ramon became close friends with Joachim Joseph, and agreed to take the Torah with him on his journey into space.

Sexton, Colleen A. *Space Shuttles*. Minneapolis, MN: Bellwether Media, 2010. Ages 4–8. This book describes how Space Shuttles carry astronauts into space and return them to Earth. Readers learn the history of the Space Shuttle, the advanced technology on board a Shuttle, and how these space vehicles perform their missions.

Shorto, Russell. *How To Fly the Space Shuttle*. Santa Fe, NM: John Muir Publications, 1992. Ages 9–12. This book explains to readers the basic principles that make spaceflight possible, as well as how Space Shuttles work and what astronauts do during a Shuttle mission.

Sloan, Peter, and Sheryl Sloan. *The Space Shuttle*. Little Littleton, MA: Sundance, 2000. This beginning reader features photographs of Space Shuttles and of life in space aboard the Space Shuttle, accompanied by text using strong word-picture correlation, large print, word counts ranging from 31 to 46, and predictable language patterns.

Software Toolworks and Follett Software Company. *The Software Toolworks Presents, Space Shuttle*. Computer game. Novato, CA: Software Toolworks, 1993. This child's game teaches users about the history of NASA and the Space Shuttle. Children playing the game make a virtual visit to Mission Control and take a virtual tour of the Space Shuttle. The game includes an experience of basic astronaut training, and players can select from several types of Shuttle mission, such as cleaning up space debris.

Space Shuttle: Find Out What's Inside. Sydney, NSW: Book Company Publishing, 2010. This "x-ray window" board book is intended for a juvenile audience. It contains pictures of the inside of a Space Shuttle and presents space travel facts.

Spangenburg, Ray, Diane Moser, and Kit Moser. *Onboard the Space Shuttle*. New York: Franklin Watts, 2002. Grades 5–9. The authors explain what life in space is like and describe the oddities that an aspiring astronaut must expect. Sections of the book cover the significant accomplishments of landmark Space Shuttle missions, such as the missions to repair the HST, as well as space stations and the *Challenger* explosion. The book also features many interesting color photographs and sidebars providing statistics, information about individual astronauts, and scientific principles related to space travel.

Steinberg, Florence S. *Aboard the Space Shuttle*. Washington, DC: NASA, 1980. This well-illustrated book is designed for classroom use. A good example of NASA's public relations material, the book is intended to familiarize children with the Shuttle and its mission.

Stille, Darlene R. *Space Shuttle*. Minneapolis, MN: Compass Point Books, 2004. Ages 4–8. This book provides a simple introduction to the Space Shuttle, describing its equipment, parts, uses, and a typical journey into space.

Stott, Carole. *Fly the Space Shuttle*. New York: Dorling Kindersley, 2001. Ages 9 and older. In this book, aimed at a juvenile audience, the reader is invited to pretend that he or she is training as an astronaut to fly on a Space Shuttle mission. The reader is introduced to other members of the crew and learns about the fun and the difficulties of living in a weightless environment, maneuvering the Shuttle orbiter in space, and deploying a satellite. The book includes an easy-to-make, three-dimensional Space Shuttle model. Pull-out and lift-the-flap sections show how the Shuttle works and how it is used.

Streissguth, Thomas. *The Challenger: The Explosion on Liftoff*. Mankato, MN: Capstone High-Interest Books, 2003. Ages 9–12. This book examines Space Shuttle *Challenger* and the events that led to its destruction, along with the effects of the disaster on NASA's space program.

Taylor, Robert. *Life Aboard the Space Shuttle*. San Diego, CA: Lucent Books, 2002. Ages 9–12. This book describes the early years of the Space Shuttle program, including the construction of the first Shuttles and the training of the crews. Chapters cover crew training, liftoff, adapting to microgravity, the problems of living in space, and how the Space Shuttle solves those problems.

Taylor, Robert. *The Space Shuttle*. San Diego, CA: Lucent Books, 2002. Ages 9–12. This book describes the history and development of the Space Shuttle, technological and political challenges to its development, and the future of the world's first reusable space vehicle.

The Space Shuttle. Videocassette (VHS). Bethesda, MD: Discovery Channel Education, 2001. This video, intended for elementary school children, looks behind the scenes at the construction and technology of the Space Shuttle and the people who make it possible.

The Space Shuttle. Videocassette (VHS). Bethesda, MD: Discovery Communications, 1996. This video, intended for grades 6–12, shows how NASA prepares the Space Shuttle for liftoff and how experiments carried out on board the Shuttle open new worlds of opportunity for scientific exploration.

Vogt, Gregory. *John Glenn's Return to Space*. Brookfield, CT: Millbrook Press, 2000. Grades 5–8. This book describes John H. Glenn Jr.'s early days as an astronaut and tells the story of his 1962 spaceflight in Friendship 7, as the first American to orbit Earth. The author recounts Glenn's return to space in 1998 at the age of 77 aboard Space Shuttle *Discovery*, emphasizing the extraordinary contrast between Glenn's first orbit around Earth and his recent Shuttle flight and using Glenn's story to symbolize the evolution of the American space program. The author also explains that Glenn joined the *Discovery* crew as a senior citizen so that he could participate in experiments aimed at measuring the effects of space travel on the human body.

Vogt, Gregory. *Space Shuttles*. Mankato, MN: Bridgestone Books, 1999. Ages 4–8. This book, full of fascinating NASA photographs to intrigue young readers, presents information about hands-on activities to promote interest in the Space Shuttle and space exploration. Labeled photodiagrams support technical vocabulary.

Walsh, Patricia, and Mark Adamic. *Space Vehicles*. Chicago: Heinemann Library, 2001. This book presents instructions for drawing the Space Shuttle, as well as other spacecraft and space vehicles, in six easy steps.

Whitehouse, Patricia. *Living in Space*. Chicago: Heinemann Library, 2004. Ages 4–8. Grades 1–3. This book gives readers a glimpse of the experiences astronauts have while living in space and on the Space Shuttle. It describes Shuttle astronauts' daily routines and explains how they receive air, water, and food, and reveals what happens to their garbage.

Whitehouse, Patricia. *Working in Space*. Chicago, IL: Heinemann Library, 2003. Ages 6 and up. This introduction to working in space discusses Space Shuttle astronaut training and different kinds of jobs in space, describing what it is like to leave Earth aboard a Space Shuttle, work in microgravity, fix satellites, work outside a Space Shuttle, and put on a spacesuit. The author also discusses what it might be like to work on other worlds. The book includes a section of space facts.

Zuehlke, Jeffrey. *The Space Shuttle*. Minneapolis, MN: Lerner, 2007. Ages 4–8. This book provides information on the Space Shuttles NASA has developed, explaining how NASA launches the Shuttle into orbit and how astronauts maneuver it in space.

Link to Part 1 (1970–1991), Chapter 15—Juvenile Literature

CHAPTER 15—*COLUMBIA* ACCIDENT AND AFTERMATH

Bea, Keith. "Disaster Relief and Responses: FY 2003 Supplemental Appropriations." CRS Report for Congress, Congressional Research Service, Library of Congress, Washington, DC, 25 August 2003. *http://assets.opencrs.com/rpts/RL31999_20030825.pdf* (accessed 11 March 2012). This report describes NASA's request to Congress for US$50 million to fund the investigation and recovery of debris from the *Columbia* disaster.

Boin, Arjen, and Paul Schulman. "Assessing NASA's Safety Culture: The Limits and Possibilities of High-Reliability Theory." *Public Administration Review* 68, no. 6 (November 2008): 1050–1062. This article reports the *Columbia* Accident Investigation Board's (CAIB's) sharp criticism of NASA's safety culture. Adopting as a benchmark the concept of a "high-reliability organization," the CAIB concluded that NASA does not possess the organizational characteristics that would have enabled it to prevent this disaster.

Cabbage, Michael and William Harwood. *Comm Check: The Final Flight of Shuttle Columbia.* New York: Free Press, 2004. Through dozens of interviews, documents, and recordings of meetings, veteran space journalists Cabbage and Harwood explore the political pressures and the management decisions that culminated in the *Columbia* accident. The authors tell the human story behind the tragic event, including the crucial errors that led to the accident and the events that unfolded in its immediate aftermath.

Chandler, David L. "Sifting Through the Clues; As the Evidence from *Columbia*'s Final Moments Piles Up, Investigators Are Faced with Several Competing Theories for What Triggered the Disaster." *New Scientist*, 15 February 2003. This article describes the efforts underway to determine the causes of the *Columbia* tragedy. Investigators had not yet discovered the cause, and the author of the article believes that the likelihood of their quickly finding a definitive answer is remote.

Chandler, David L. "'There Was Zero We Could Have Done . . .'; Or Was There? If NASA Had Known There Was No Prospect of *Columbia* Landing Safely, Could It Have Rescued the Crew?" *New Scientist*, 22 March 2003. This article reports that, shortly after *Columbia* broke up on reentry, Space Shuttle Program Manager Ronald D. Dittemore said that NASA had been aware that a piece of foam insulation had fallen from the Shuttle's external tank during liftoff. In spite of this, NASA's Mission Control decided not to photograph the orbiter's underside to determine whether its heat-protection tiles were damaged. The article speculates about what NASA could have done to prevent the tragedy if it had discovered the damage before *Columbia*'s return to Earth.

Chien, Philip. *Columbia—Final Voyage: The Last Flight of NASA's First Space Shuttle.* Chichester, UK: Praxis Publishing, 2006. This book explains STS-107, *Columbia*'s final mission, a "free flyer" mission in which NASA planned for the crew to spend 16 days in orbit, performing dozens of scientific experiments. The book devotes one chapter to each STS-107 astronaut. The author criticizes the media for covering the mission only after the catastrophe had occurred.

Cole, Michael D. *The Columbia Space Shuttle Disaster: From First Liftoff to Tragic Final Flight*. Rev. ed. Berkeley Heights, NJ: Enslow Publishers, 2003. This book contains technical details about the Space Shuttle, as well as personal information about the astronauts aboard *Columbia* on its final mission, STS-107. Photographs introduce young readers to the crew and help them understand the events that led to the tragic loss of the spacecraft and its crew.

Columbia Accident Investigation Board. *Report*. 6 vols. Washington, D.C.: Government Printing Office, 2003. *http://permanent.access.gpo.gov/lps39093/lps39093/caib.nasa.gov/news/ report/volume5/default.html* (accessed 9 March 2012). The *Columbia* Accident Investigation Board (CAIB) wrote this report as a framework for national debate about the future of human spaceflight. The CAIB recommends expediting the replacement of the Space Shuttle, the primary means for transporting humans to and from Earth's orbit. In the final report, the Board makes 29 recommendations for NASA, including 15 that it considers essential prerequisites for the Shuttles' safe return to flight. The report also concludes that NASA lacks a strong safety culture.

"Columbia Investigation." *Spaceflight*, September 2003. This article describes ongoing efforts in the investigation of the Space Shuttle *Columbia* tragedy. The article notes that the *Columbia* Accident Investigation Board (CAIB) has delayed delivery of its accident report. The CAIB has warned NASA about potential problems with the bolts that connect the solid rocket boosters to the Space Shuttle's external tank, indicating that changes to the tanks will be necessary. In addition, the article predicts a radical reorganization of NASA's management.

Covault, Craig. "Columbia Probe Shifts." *Aviation Week & Space Technology*, 21 April 2003. The author of this article explains that the *Columbia* Accident Investigation Board (CAIB) is focusing on a possible scenario to explain the accident, in which a fracture of a wing T-seal opened a 1-inch-wide (0.6-centimeter-wide) vertical slit in the orbiter's leading edge. The Board is also examining the fracture of a reinforced-carbon-carbon panel.

Covault, Craig. "Columbia Revelations." *Aviation Week & Space Technology*, 3 March 2003. The author focuses on alarming e-mails among NASA engineers and contractor personnel in the days before *Columbia*'s disastrous reentry. The e-mail exchange provides deepening evidence that serious concerns at NASA about *Columbia*'s left wheel well and the survivability of its left wing were widespread and growing during the final days of the flight.

Covault, Craig. "Critical Columbia Tests." *Aviation Week & Space Technology*, 17 March 2003. This article describes efforts by the *Columbia* Accident Investigation Board (CAIB) and NASA to lead a critical series of debris-impact tests. The tests would help determine whether a large piece of external tank foam could have fatally damaged *Columbia* or whether other factors, such as materials degradation of the aging spacecraft, could have played a pivotal role in the tragedy.

Covault, Craig. "Growing Evidence Points to Columbia Wing Breach." *Aviation Week & Space Technology*, 17 February 2003. This article reports on the growing evidence that wing failure played a large part in the *Columbia* disaster. The author points out that sensors in the wheel well and on the trailing edge of the wing provided some of the first signs of trouble.

Covault, Craig. "NASA's Eroding Safety." *Aviation Week & Space Technology*, 12 May 2003. Covault predicts that the *Columbia* Accident Investigation Board's (CAIB's) report will cite serious deficiencies in NASA's overall safety program as a root cause or significant contributing factor to the loss of Space Shuttle *Columbia* and its crew. The author anticipates that the report will question whether NASA's oversight of the spacecraft was effective, once NASA had transferred specific duties of vehicle quality control to the United Space Alliance, as required under the Space Flight Operations Contract.

Covault, Craig. "New Shuttle Concerns Aired." *Aviation Week & Space Technology*, 15 July 1996. The author reports that the White House, NASA, and the aerospace industry are assessing the effect of an independent Aerospace Safety Advisory Panel (ASAP) report that raises serious Space Shuttle safety issues, including the potential for increased risk of a Shuttle accident. The concerns stem from NASA's plan to reduce its costs by shifting additional operational responsibilities to the commercial contractor United Space Alliance. NASA and Thiokol are in the midst of a major investigation of the Space Shuttle's solid rocket motors to determine why hot gas penetrated into new areas of all six field joints on the two boosters that launched the orbiter *Columbia*.

Covault, Craig. "Roaring Comeback." *Aviation Week & Space Technology*, 11 July 2005. This article discusses the Space Shuttle main engines (SSMEs) and solid rocket boosters (SRBs) that will propel *Discovery*'s launch into orbit on its return-to-flight mission. In the two years since the *Columbia* accident, NASA's SSME and SRB programs have substantially increased the rigor of their testing and quality oversight.

Covault, Craig. "Rough Wing + Debris + A Fatal Combination?" *Aviation Week & Space Technology*, 24 February 2003. This article presents information about Shuttle *Columbia*'s left wing and the possibility that damage to the wing led to *Columbia*'s destruction. NASA's Johnson Space Center has documents from as early as 1988 indicating that wing roughness similar to that associated with *Columbia*'s left wing could result in a catastrophic burn-through when combined with potentially significant impact damage resembling the damage that occurred to the left wing in the 1 February reentry accident.

Covault, Craig. "Shuttle Shaping Up; The Shuttle's Pace Toward Return-to-Flight Is Accelerating in Spite of Orbital Repair Challenges." *Aviation Week & Space Technology*, 13 December 2004. This article claims that, following the *Columbia* disaster, NASA may take as long as two years to build a fully certified thermal protection system and to develop the capability for in-orbit repair of the Shuttle *Columbia*'s wing leading edge. However, the safety tradeoffs are positive, overall, enabling return-to-flight preparations to accelerate this month and making a return of the Space Shuttle to space in the spring a real possibility.

Covault, Craig. "Tank Debris Assessment Spotted 'No Safety Issue'." *Aviation Week & Space Technology*, 10 February 2003. This article describes the calculations performed by NASA and its contractors to assess the danger of damage to *Columbia* from foam falling from the external tank. The postlaunch analysis of *Columbia* determined that foam insulation debris posed no serious threat to the safety of the orbiter and crew.

Covault, Craig. "USAF Imagery Confirms Columbia Wing Damaged." *Aviation Week & Space Technology*, 10 February 2003. This article describes images taken of *Columbia* as it approached Earth for the last time. The images show serious structural damage to the inboard leading edge of *Columbia*'s left wing. The author speculates about what the images can tell investigators about the accident.

Cowen, Ron. "Columbia Disaster." *Science News*, 8 February 2003. The author claims that Space Shuttle *Columbia*, which broke apart on 1 February 2003 just minutes before it was scheduled to land, may have been doomed at liftoff. *Columbia* sustained damage during liftoff when foam detached from the Shuttle's external tanks and struck the orbiter's left wing.

Deal, Duane W. "Beyond the Widget; Columbia Accident Lessons Affirmed." *Air & Space Power Journal* 18, no. 2 (22 June 2004): 31. According to the author of this article, the *Columbia* Accident Investigation Board (CAIB) has pointed out several lessons learned from the *Columbia* disaster, and senior leaders of other high-risk operations should seek to learn from the CAIB's conclusions. For example, managers of such operations should encourage dissenting opinions and should maintain checks and balances within their organizations.

Evans, Ben. *Space Shuttle Columbia: Her Missions and Crews*. Chichester, UK: Praxis Publishing, 2005. This book, written by the scientists and researchers who developed and supported *Columbia*'s many payloads, the engineers who worked on the spacecraft, and the astronauts who flew it, comprises detailed descriptions of Space Shuttle *Columbia*'s 28 missions. The book is intended as a tribute to *Columbia* and to the people who have supported the Shuttle program.

"Excerpts from Joint Congressional Hearing on Loss of Shuttle Columbia." *New York Times*, 13 February 2003 Witnesses testifying before a joint congressional hearing into the loss of Space Shuttle *Columbia* included NASA Administrator Sean O'Keefe, U.S. Senators Olympia J. Snowe (R-ME) and Clarence William "Bill" Nelson (D-FL), and U.S. Representative Sherwood Boehlert (R-NY).

"Excerpts from NASA E-mails About Space Shuttle Before It Disintegrated." *New York Times*, 27 February 2003. This article contains excerpts from e-mail messages exchanged among engineers during Space Shuttle *Columbia*'s flight. NASA released these e-mails in February 2003.

Fahey, Kathleen. *Challenger and Columbia*. Milwaukee, WI: Gareth Stevens Publishing, 2005. Ages 9 and older. The book describes the events leading up to the loss of Space Shuttles

Challenger and *Columbia*. Based on detailed descriptions, expert testimony, and firsthand accounts, the author takes an in-depth look at the *Challenger* and *Columbia* tragedies and discusses the lessons learned from them.

Furniss, Tim. "Columbia Tragedy Piles Pressure on Space Program." *Spaceflight*, March 2003. This short news article describes how NASA reported to the media the sequence of events that occurred during Space Shuttle *Columbia*'s reentry. The article explains that *Columbia* was not equipped to perform a self-check for damage before returning to Earth.

Furniss, Tim. "Investigators Close In on Causes of Columbia Loss." *Spaceflight*, May 2003. This article reviews the findings of the investigators regarding the *Columbia* accident. The author claims that investigators have identified the series of events that led to the disintegration of *Columbia*. The article reviews the possibility that the Shuttle's "black box" flight data recorder will shed light on the tragedy and discusses the efforts to retrieve Shuttle debris.

Garrett, Terence M. "Whither Challenger, Wither Columbia." *American Review of Public Administration* 34, no. 4 (December 2004): 389–402. The *Columbia* Accident Investigation Board (CAIB) scrutinized similarities in NASA management's decision making in the ill-fated missions of Space Shuttles *Challenger* and *Columbia*, determining that NASA's organizational and management culture was a key factor in the loss of both orbiters. Both tragedies occurred after senior-level managers ignored advice from experts within the NASA organization. However, in this article, the author argues that NASA's management culture is not the sole cause of the tragedies, only a contributing factor.

Geenty, John. "Flights of Fancy." *Spaceflight*, January 2005. This article describes how NASA planned its Space Shuttle missions before the *Columbia* tragedy. Originally, NASA had planned for the Shuttle to fly more often, with a greater range of payloads. However, the losses of *Challenger* and *Columbia* led to a change in Shuttle flights and schedules. Moreover, the loss of *Columbia* meant that future Shuttle missions would have to support the ISS, with no flights to spare for science programs.

Hall, Joseph Lorenzo. "Columbia and Challenger: Organizational Failure at NASA." *Space Policy* 19, no. 4 (November 2003): 237–247. This article outlines some of the critical features of NASA's organization. The author discusses organizational change at NASA, namely "path dependence" and "normalization of deviance." He reviews the reasons that experts have called the *Challenger* tragedy an organizational failure. Finally, he argues that the recent *Columbia* accident also displays characteristics of organizational failure and proposes recommendations for the future.

Hosenball, Mark, Jerry Adler, Anne Belli Gesalman, and Tamara Lipper. "Falling to Earth." *Newsweek*, 17 February 2003. This article describes the investigation into the explosion of *Columbia* and the Shuttle debris found throughout Texas and Louisiana. The author remarks that some journalists have denounced the administration of President George W. Bush for wasting money on showy but scientifically trivial unpiloted spaceflights, while others have condemned the Bush administration for not spending enough. In addition, the author reports NASA engineers' ideas about what might have gone wrong on *Columbia*.

Jenkins, Dennis R., and Jorge R. Frank. *Return to Flight Space Shuttle Discovery: Photo Scrapbook*. North Branch, MN: Specialty Press, 2006. This book shows photographs from the launch and flight of STS-114, the Space Shuttle program's return to flight following the *Columbia* disaster. This flight was the most photographed flight ever, with numerous cameras on ships, on aircraft, and on the ground tracking the vehicle during ascent, and the crew of the ISS taking a series of detailed photographs as *Discovery* approached the ISS. In addition, the crew of *Discovery* used cameras in the cockpit and on a long, robotic arm to examine almost every inch of *Discovery*.

Klerkx, Greg. "It's Back . . .; And This Time There Is More Than Ever Riding on the Space Shuttle's Success." *New Scientist*, 30 April 2005. This article describes NASA personnel's anticipation of the return to flight of Space Shuttle *Discovery* after the *Columbia* disaster. The article includes a checklist of the eight items—out of the list of the *Columbia* Accident Investigation Board's (CAIB's) 15 recommendations—that NASA completed before returning the Space Shuttle to flight.

Klesius, Michael. "The Evolution of the Space Shuttle." *Air & Space*, July 2010. In this article, the author discusses the history of the Space Shuttle program, explores modifications to the design of the Space Shuttle considered throughout the program, and examines the errors leading to the loss of *Challenger* in January 1986 and *Columbia* in February 2003.

Klotz, Irene. "Crunch Time for Shuttle As Safety Demands Pile Up." *New Scientist*, 15 July 2006. The author describes NASA's reactions to Space Shuttle *Discovery*'s return-to-flight mission. Instead of experiencing jubilation, NASA personnel are bracing for the effort they must make to complete the ISS before the Shuttle's retirement in 2010.

Kluger, Jeffrey, Cathy Booth, Matthew Cooper, Sally B. Donnelly, Deborah Fowler, Hilary Hylton, Broward Liston, and David E. Thigpen. "Fragments of a Mystery." *Time*, 17 February 2003. The article reports that NASA investigators have determined that a loose chunk of insulating foam damaged Space Shuttle *Columbia*'s skin and led to its breakup in the skies over Texas during reentry. The authors found old NASA memoranda warning of the possibility of this type of accident.

Kluger, Jeffrey, Stefano Coledan, Deborah Fowler, and Eric Roston. "Why NASA Can't Get It Right." *Time*, 1 August 2005. This article reports that four pieces of insulating foam— including one as large as a skateboard—have spun off Space Shuttle *Discovery*'s external fuel tank during liftoff. This is the same type of debris that damaged *Columbia*'s wing and doomed the craft. Photographs of the Shuttle have shown at least 25 dings in *Discovery*'s insulating tiles, including a 1.5-inch (3.8-centimeter) divot near the nose.

Kortenkamp, Steve. *Space Shuttles*. Mankato, MN: Capstone Press, 2008. Ages 4–8. This book describes the history and uses of NASA's Space Shuttles, including both the practical uses and the drawbacks of reusable Space Shuttles. The author explains how the Shuttle flies and how it launches satellites. In addition, he describes how astronauts travel into space to make repairs on missions such as the missions to service the HST. The book also includes information about the two Shuttle disasters.

Kosova, Weston, Sam Seibert, Seth Mnookin, and Joshua Hammer. "The Right Stuff." *Newsweek*, 10 February 2003. This article profiles the astronauts who perished on *Columbia*: Richard D. Husband, William C. McCool, Michael P. Anderson, Kalpana Chawla, David M. Brown, Laurel B. Clark, and Ilan Ramon.

Kruger, Jeffrey, Missy Adams, Cathy Booth-Thomas, John F. Dickerson, Sally Donnelly, Deborah Fowler, Greg Fulton, Jerry Hannifin, Eric Roston, Elaine Shannon, Mark Thompson, Karen Tumulty and Doug Waller.. "What Went Wrong?" *Time*, 28 July 2005. This article describes efforts to determine the cause of the explosion of Space Shuttle *Columbia* upon reentry to the Earth's atmosphere. The author explains how a Shuttle crew prepares for reentry and describes the items that the investigators will examine during the inquest into the *Columbia* accident.

Lawler, Andrew. "Vision, Resources in Short Supply from Damaged U.S. Space Program." *Science* 301, no. 5638 (5 September 2003): 1300–1303. The author indicates that politicians in Washington, DC, need to respond to the *Columbia* Accident Investigation Board (CAIB) report issued on 26 August 2003 and to provide leadership. The author believes that the report compels Congress and the president to provide clearer vision and more substantial funding, to support a robust human space exploration program.

Lerner, Preston. "NASA's Fixer-Upper Flies Again: Two Years After Columbia, NASA Is Counting on a Refurbished Space Shuttle To Revive Its Floundering Human-Spaceflight Program." *Popular Science*, May 2005. This article describes the May 2005 launch of STS-114, the return-to-flight Space Shuttle mission, using the refurbished Space Shuttle *Discovery*.

"Lost in Space." *Economist*, 30 August 2003. This article focuses on key controversies surrounding the Space Shuttle *Columbia* accident. One of the more troubling issues discussed in the article is the possibility that another Shuttle—*Atlantis*—might have undertaken a rescue mission if the mission's managers had recognized that the foam had damaged *Columbia*.

MacRae, Duncan. "Columbia Break-up Turns Focus on New-Generation RLV." *Interavia Business & Technology*, February 2003. This article describes the effect of the loss of *Columbia* on future Space Shuttle flights, providing details of the *Columbia* disaster. The author discusses the need to improve existing production capability, to enable the construction of additional space vehicles and the development of second-generation space vehicles.

MacRae, Duncan. "Columbia Report Blasts NASA." *Interavia Business & Technology*, September 2003. This article reviews the official results of the investigation of the Space

Shuttle *Columbia* accident conducted by the *Columbia* Accident Investigation Board (CAIB), presenting the physical and organizational causes of the accident.

Martin, Ryan M, and Louis A. Boynton. "From Liftoff to Landing: NASA's Crisis Communications and Resulting Media Coverage Following the Challenger and Columbia Tragedies." *Public Relations Review* 31, no. 2 (June 2005): 253–261. NASA's public relations effort following the explosion of *Challenger* in 1986 is considered an example of crisis communications failure. However, after the *Columbia* disaster in 2003, NASA was praised for its successful handling of the crisis. Using widely accepted crisis communication concepts associated with stakeholder theory, the author discusses how four newspapers presented NASA's crisis communication efforts following the two crises, showing that the print media accorded NASA more positive coverage following the *Columbia* disaster than the *Challenger* disaster.

Morring, Frank, Jr. "Connecting the Dots." *Aviation Week & Space Technology*, 14 April 2003. This article examines NASA's efforts to correct the most likely cause of the *Columbia* tragedy before the *Columbia* Accident Investigation Board (CAIB) writes its report. NASA hopes to begin making corrections so that the Space Shuttles will be cleared to resume flights to the ISS as soon as possible. The article also notes the CAIB's efforts to recover *Columbia* debris and to use wind tunnel exercises to determine the cause of the tragedy.

Morring, Frank, Jr. "Shuttle Accident Puts 'Everything on Table'." *Aviation Week & Space Technology*, 10 February 2003. This article describes the efforts of Congress and the Bush administration to determine the cause of the *Columbia* accident. The congressional inquiry covered technical issues and political issues underlying space policy decisions on matters such as construction of the ISS, the need for RLVs, and funding to ensure Space Shuttle safety. The accident halted NASA's policy reviews regarding the final configuration of the ISS and plans to build an Orbital Space Plane.

Mowbray, Scott. "After Columbia: The ISS in Crisis." *Popular Science*, April 2003. This article focuses on how Space Shuttle *Columbia*'s accident has affected the construction of the ISS. The author explains that NASA had intended 2003—the year of the *Columbia* tragedy—as a pivotal year for building the space station. Instead, NASA will have to work with Russia's space authorities to ensure the safe operation of the ISS during the long grounding before another Shuttle is cleared to return to space.

Murnane, Andrew W. "Theft of Debris from the Space Shuttle Columbia: Criminal Penalties." CRS Report for Congress, Congressional Research Service, Library of Congress, Washington, DC, 12 June 2003. *http://www.fas.org/spp/civil/crs/RS21417.pdf* (accessed 8 March 2012). This report explains why recovery of Space Shuttle *Columbia* debris was considered vital to the investigation into *Columbia*'s final moments of flight. The report also describes possible criminal penalties for theft of government property. According to the report, at least six individuals were indicted in Texas and Florida on charges that they stole *Columbia* debris.

Murnane, Andrew W., and Daniel Inkelas. "Liability Issues Associated with the Space Shuttle Columbia Disaster." CRS Report for Congress, Congressional Research Service, Library of Congress, Washington, DC, 12 February 2003. *http://www.fas.org/spp/civil/crs/ RS21426.pdf* (accessed 8 March 2012). This report describes legal principles and processes that govern possible compensation for loss of life and property resulting from the Space Shuttle *Columbia* disaster.

NASA. Johnson Space Center. *STS-107 Memories*. Washington, DC: U.S. Government Printing Office, 2006. CD. This CD, commemorating the seven crew members who lost their lives on Space Shuttle *Columbia* in February 2003, features biographies and photographs of the crew as they worked and trained together.

NASA. Return to Flight Task Group. *Return to Flight Task Group—Final Report July 2005*. Washington, DC, 2005. *http://www.nasa.gov/pdf/125343main_RTFTF_final_081705.pdf* (accessed 8 March 2012). Soon after publication of the *Columbia* Accident Investigation Board (CAIB) report, NASA Administrator Sean O'Keefe appointed a Return to Flight (RTF) Task Group to assess NASA's progress implementing CAIB recommendations before resuming Shuttle flights. Relative to the 15 specific recommendations that the CAIB indicated should be implemented before returning to flight, NASA has met or exceeded 12. The remaining three recommendations were so challenging that NASA could not comply completely with the intent of the CAIB, but conducted extensive study, analyses, and hardware modifications that resulted in substantial progress toward making the vehicle safer.

Oberg, James. "Puncture Repair Kit; The Shuttle Columbia Might Have Been Saved If the Crew Had Been Able To Fix A Hole While in Orbit." *New Scientist*, 15 November 2003. This article examines NASA's efforts to create patch kits that astronauts could use to make repairs to orbiters if they sustained damage such as *Columbia* experienced during its last takeoff. Besides trying out NASA's repair kit, the author identifies other options for inflight repairs to external surfaces of the spacecraft.

Pianin, Eric, and Kathy Sawyer. "Skeptics Say Shuttle Worn Out, Obsolete." *Washington Post*, 12 May 2003. As NASA recovers from the *Columbia* disaster and works on returning the three remaining Space Shuttles to flight, a growing chorus of skeptics says that the 20-year-old spacecraft may have become unacceptably worn out and obsolete.

Powell, Joel W. "Columbia Investigation: Piecing Together the Evidence." *Spaceflight*, October 2003. This article describes the efforts made to find clues among the debris from the *Columbia* tragedy. Investigators recovered the Orbiter Experiments recorder box and performed forensic analysis on the debris. The article contains photographs of the debris.

"Radical Shuttle Changes in Light of Columbia Inquiry." *Spaceflight*, August 2003. This article predicts that the *Columbia* Accident Investigation Board (CAIB) will recommend changes to the Space Shuttle program. The article discusses the possibility that NASA could have launched Space Shuttle *Atlantis* to rescue the *Columbia* crew if NASA had known of the impending dangers to *Columbia* of reentry. Changes to NASA's quality control and assurance programs and inspection programs may be required.

Reichhardt, Tony. "Columbia Explosion May Trigger Fatal Delays for Space Station." *Nature* 421, no. 6923 (6 February 2003): 561. This article focuses on the effect of Space Shuttle *Columbia*'s accident on the ISS program and on NASA's other projects.

Reichhardt, Tony. "NASA: Trawling Through the Wreckage." *Nature* 426, no. 6968 (18 December 2003): 754–755. This article reports on the causes of the accident that destroyed Space Shuttle *Columbia*. Results of the NASA investigation indicate that insulating foam could have punched a hole in the Shuttle.

Return to Flight Space Shuttle. Denver, CO: Denver Museum of Nature and Science, 2006. DVD. This DVD program describes NASA's efforts to prepare STS-114, the return-to-flight mission following the Space Shuttle *Columbia* disaster, as well as NASA's preparations for STS-121, which tested the safety and repair techniques imposed after the *Columbia* disaster.

Schuiling, Roelof. "STS-107: Columbia's Final Mission." *Spaceflight*, April 2003. This article discusses Space Shuttle *Columbia*'s final mission, STS-107, describing launch preparation, payloads, crew members, and daily activities during the 15-day mission to conduct scientific experiments. *Columbia* and its crew were lost during reentry over east Texas on 1 February 2003, at about 9:00 a.m. (EST).

Schwartz, John. "Shuttle Surface More Vulnerable Than Suspected." *New York Times*, 20 January 2005. In interviews this week, NASA officials reported the results of impact tests and analyses performed as part of the return-to-flight effort after the 2003 *Columbia* disaster. The tests showed that pieces of insulating foam weighing less than half an ounce could cause small cracks and damage to the surface coating on the heat-resistant panels on the leading edge of the wing.

Scott, William B. "Playing the Odds with Space Debris." *Aviation Week & Space Technology*, 17 February 2003. NASA investigation teams reassessed potential risks of orbital debris to Space Shuttle tiles and reinforced carbon-carbon leading edges after U.S. Air Force radar reported spotting an object after *Columbia* reached orbit. The report prompted renewed scrutiny of "space junk" and how it might have damaged the Shuttle's thermal protection system.

Seife, Charles. "Columbia Disaster Underscores the Risky Nature of Risk Analysis." *Science* 299, no. 5609 (14 February 2003): 1001. This article focuses on NASA's risk analysis of the probability of losing a Space Shuttle and its crew. The author believes that the use of probabilistic risk assessment (PRA) would improve NASA's risk analysis. PRA is a method engineers use to determine the likely failure rate of a complex system, such as a Space Shuttle, and to pinpoint the elements most likely to contribute to such a failure.

Shiner, Linda. "The Space Shuttle Returns." *Air & Space*, May 2005. The author provides an update from NASA's Michoud Assembly Facility about construction of Space Shuttle *Discovery*. The article also describes the cause of the disintegration of Space Shuttle *Columbia* on 1 February 2003; results of the investigation conducted by engineers from NASA's Johnson Space Center in Houston, Texas, into the damaged panels of insulation; and reasons for the changes in the launch schedule of the Space Shuttle.

Smith, Marcia S. "NASA's Space Shuttle Columbia: Quick Facts and Issues for Congress." CRS Report for Congress, Congressional Research Service, Library of Congress, Washington, DC, 13 February 2003. *http://assets.opencrs.com/rpts/RS21408_20030213.pdf* (accessed 8 March 2012). This report includes facts about Space Shuttle *Columbia* and the investigation into the loss of *Columbia*, as well as a presentation to Congress about space program issues.

Smith, Marcia S. "NASA's Space Shuttle Columbia: Synopsis of the Report of the *Columbia* Accident Investigation Board." CRS Report for Congress, Congressional Research Service, Library of Congress, Washington, DC, 2 September 2003. *http://assets.opencrs. com/rpts/RS21606_20030902.pdf* (accessed 8 March 2012). This report presents a brief synopsis of the *Columbia* Accident Investigation Board's (CAIB's) report. The CAIB specified that 15 out of its 29 recommendations must be completed before the Shuttle's return to flight. The synopsis includes reasons for the accident, a discussion about whether NASA should have taken pictures of *Columbia* before its return to Earth, and speculation about whether NASA could have saved the crew.

Smith, Marcia S. "NASA's Space Shuttle Program: Issues for Congress Related to the Columbia Tragedy and 'Return to Flight'." CRS Report for Congress, Congressional Research Service, Library of Congress, Washington, DC, 2 June 2005. *http://fpc.state.gov/ documents/organization/48804.pdf* (accessed 7 March 2012). This report describes Space Shuttle *Discovery*'s 9 August 2005 return-to-flight (RTF) mission, designated STS-114, the first Shuttle launch after the *Columbia* tragedy. On 27 July 2005, NASA announced that it planned to postpone indefinitely the second RTF mission, because a problem had occurred during *Discovery*'s launch that was similar to the event that led to the loss of *Columbia*.

Smith, Marcia S. "NASA's Space Shuttle Program: The Columbia Tragedy, the Discovery Mission, and the Future of the Shuttle." CRS Report for Congress, Congressional Research Service, Library of Congress, Washington, DC, 4 January 2006. *http://www. fas.org/sgp/crs/space/RS21408.pdf* (accessed 7 March 2012). This report describes the *Columbia* accident investigation and *Discovery*'s return-to-flight (RTF) mission, designated STS-114, and provides Congress with information about future Shuttle flights.

Smith, Marcia S. "National Aeronautics and Space Administration: Overview, FY 2004 Budget in Brief, and Issues for Congress." CRS Report for Congress, Congressional Research Service, Library of Congress, Washington, DC, 23 June 2003. *http://assets.opencrs.com/ rpts/RS21430_20030728.pdf* (accessed 12 March 2012). This report focuses on NASA's US$15.5 billion FY 2004 budget request. The author discusses the investigation of the

Space Shuttle *Columbia* tragedy on 1 February 2003 and its implications for NASA and for the space program as a whole.

Smith, Marcia S., and Daniel Morgan. "The National Aeronautics and Space Administration: Overview, FY 2005 Budget in Brief, and Key Issues for Congress." CRS Report for Congress, Congressional Research Service, Library of Congress, Washington, DC, 5 October 2004. *http://www.fas.org/spp/civil/crs/RS21744.pdf* (accessed 8 March 2012). This report describes NASA's FY 2005 budget request for US\$16.2 billion, a 5.6 percent increase over its FY 2004 appropriation of US\$15.4 billion. The report also discusses the new space exploration goals that President George W. Bush announced on 14 January 2004, NASA's plans to return the Space Shuttle to flight status following the Space Shuttle *Columbia* accident, and the health of NASA's aeronautics research program.

Smith, R. Jeffrey, and Joe Stephens. "Safety an Issue Since '90s; Experts Critical of Shuttle Program's Budget Cuts." *Washington Post*, 3 February 2003. Despite the objections of many safety experts, NASA chose to reduce its costs by transforming the Space Shuttle program from a largely government-run effort to a program in which private contractors received more than 90 percent of its funds and operated under the supervision of only a few hundred full-time government employees. In the wake of Saturday's loss of one of the four Space Shuttles, *Columbia*, analysts are freshly scrutinizing NASA's decision.

Space Shuttle a Remarkable Flying Machine. Houston, Texas: TaLas Enterprises, 1994. Videocassette (VHS). This video profiles Space Shuttle *Columbia* from launch, to its activities in orbit, to landing. *Columbia* was the first reusable, fixed-wing spacecraft to go into orbit.

"STS-107 Crew Profiles." *Spaceflight*, April 2003. This one-page article describes the seven crew members of STS-107, who perished in the *Columbia* tragedy, providing brief biographical information and a picture of each crew member.

Tompkins, Phillip K. *Apollo, Challenger, Columbia: The Decline of the Space Program: A Study in Organizational Communication*. New York: Oxford University Press, 2004. This book portrays NASA from 1958 to 2003, concentrating on several specific points in the history of the space program, including 1986 and 2003, the periods surrounding the two Space Shuttle disasters. The author investigates the internal communication failures that resulted in the 1986 explosion of *Challenger* and the catastrophic failure of *Columbia* in 2003.

U.S. Congress. House of Representatives. Committee on Science. *NASA's Response to the Columbia Report*. 108th Cong., 1st sess., 10 September 2003. *http://www.gpo.gov/fdsys/pkg/CHRG-108hhrg89217/pdf/CHRG-108hhrg89217.pdf* (accessed 30 March 2012). In testimony at this hearing, NASA Administrator Sean O'Keefe and *Columbia* Accident Investigation Board (CAIB) chair Harold W. Gehman Jr. review the CAIB report findings and recommendations on Space Shuttle *Columbia*'s 1 February 2003 accident and provide an overview of NASA's response to the report. The volume includes the text of NASA's Implementation Plan for Return to Flight and Beyond.

U.S. Congress. House of Representatives. Committee on Science. Subcommittee on Space and Aeronautics. *Space Shuttle Safety*. 106th Cong., 1st sess., 23 September 1999. *https://web. lexis-nexis.com/congcomp/* (accessed 11 November 2011 via Proquest Congressional). Following the electrical wiring problems that occurred during the July 1999 mission of *Columbia*, STS-93, this hearing examines safety issues of the Space Shuttle program.

U.S. Congress. Joint Hearing of the House Committee on Science, Subcommittee on Space and Aeronautics, and the Senate Committee on Commerce, Science, and Transportation. *Space Shuttle Columbia*. 108th Cong., 1st sess., 12 February 2003. *http://www.gpo.gov/ fdsys/pkg/CHRG-108hhrg85090/pdf/CHRG-108hhrg85090.pdf* (accessed 30 March 2012). This joint hearing reviews the events surrounding the loss of Space Shuttle *Columbia* on 1 February 2003 and examines NASA's investigation into the accident.

U.S. Congress. Senate. Committee on Commerce, Science, and Transportation. *NASA's Space Shuttle Program*. 108th Cong., 2nd sess., 8 September 2004. *http://www.gpo.gov/fdsys/ pkg/CHRG-108shrg36172/pdf/CHRG-108shrg36172.pdf* (accessed 30 March 2012). This hearing examines the status of NASA's Space Shuttle program return-to-flight efforts in the aftermath of the loss of Space Shuttle *Columbia* on 1 February 2003 and assesses NASA's progress implementing the *Columbia* Accident Investigation Board's (CAIB's) recommendations in response to the accident.

U.S. Government Accountability Office. "Maritime Security: Better Planning Needed To Help Ensure an Effective Port Security Assessment Program." Report no. GAO-04-1062, Washington, DC, September 2004. *http://www.gao.gov/new.items/d041062.pdf* (accessed 7 March 2012). This report about the Port Security Assessment Program includes a map of Texas showing the debris field from the *Columbia* disaster.

Van der Haar, Gerard. "Columbia Inquiry." *Spaceflight*, April 2009. This article describing the inquiry into the Space Shuttle *Columbia* tragedy includes an image of *Columbia* taken from Kirtland Air Force Base, indicating a problem with the orbiter's left wing. The article describes NASA's accident reports and investigations and *Columbia* Accident Investigation Board (CAIB) activities.

"Where Now for NASA?" *Nature* 421, no. 6923 (6 February 2003): 559. This article discusses the implications of the 2003 loss of Space Shuttle *Columbia* and its crew, offering an opinion about how the tragedy will affect other space exploration projects.

White, Thomas Gordon, Jr. "The Establishment of Blame as a Framework for Sensemaking in the Space Policy Subsystem: A Study of the Apollo 1 and Challenger Accidents— OpenThesis." PhD Dissertation, Virginia Polytechnic Institute and State University, Blacksburg, VA, 2000. *http://www.openthesis.org/documents/Establishment-Blame-As-Framework-Sensemaking-582553.html* (accessed 8 March 2012). This dissertation investigates how the establishment of blame becomes a framework for sensemaking in a national policy subsystem. Using the Apollo 1 and Space Shuttle *Challenger* accidents as case studies, the author examines the space policy subsystem's response to these two accidents and the process of establishing culpability.

See also Part 1 (1970–1991), Chapter 7—Challenger Accident and Aftermath

CHAPTER 16—THE SPACE SHUTTLE AND THE *MIR* SPACE STATION

Begley, Sharon, Ginny Carroll, Peter Katel, Catharine Skipp, and Peter Annin. "Down to Earth." *Newsweek*, 7 October 1996. This article profiles astronaut Shannon W. Lucid, examining her record 188 days in space (179 days aboard *Mir* and nine days total on Space Shuttle trips to and from *Mir*) and her health after returning to Earth in September 1996. The authors also discuss Lucid's education and her career background.

Bredeson, Carmen. *Shannon Lucid Space Ambassador*. Brookfield, CT: Millbrook Press, 1998. *http://www.barnesandnoble.com/s/Shannon-Lucid-Space-Ambassador-?keyword= Shannon+Lucid+Space+Ambassador.&store=ebook* (accessed 8 March 2012). This electronic book chronicles the life of astronaut Shannon W. Lucid, from her childhood in Oklahoma, through her various Space Shuttle missions, to her six months aboard the *Mir* space station.

Covault, Craig. "Shuttle Delay Forces Changes in U.S.-Russian Operations." *Aviation Week & Space Technology*, 22 July 1996. This article explores NASA's decision to delay the launch of STS-79 to *Mir* by six weeks because of problems with the Space Shuttles' solid rocket boosters, a decision that will affect Shuttle and *Mir* operations well into the following year. NASA had planned for U.S. astronaut Shannon W. Lucid, who is currently aboard *Mir*, to spend 4.5 months at the Russian space station, but this delay will force her to remain approximately six weeks longer than planned.

Cowen, Ron. "The International Approach." *Science News*, 20 May 1995. The author reviews the plans to link Space Shuttle *Atlantis* with the Russian space station *Mir*. This maneuver will lay the groundwork for construction of an international space station.

Foale, Colin. *Waystation to the Stars: The Story of Mir, Michael, and Me*. London: Headline Book Publishing, 2000. This book, written by the father of British-American astronaut C. Michael Foale, describes the astronaut's activities and experiences on *Mir* for five months in 1997. During his stay on *Mir*, Foale had to deal with fire, collision, and computer failure.

Kanas, Nick, Vyacheslav Salnitski, Ellen M. Grund, Vadim Gushin, Daniel S. Weiss, Olga Kozerenko, Alexander Sied, and Charles R. Marmar. "Social and Cultural Issues During Shuttle/*Mir* Space Missions." *Acta Astronautica* 47, no. 2–9 (November 2000): 647–655. This article summarizes the findings from a NASA-funded study conducted during several Shuttle-*Mir* space missions. The study related to social and cultural issues involving the American and Russian Shuttle and *Mir* crew and Mission Control staff.

Kidger, Neville. "STS-71 Preview." *Spaceflight*, June 1995. This article previews STS-71, the mission of Space Shuttle *Atlantis* to link up with *Mir*. The article discusses NASA's preparations for *Atlantis*'s mission, describes the Orbiter Docking System, and explains the docking maneuvers that *Atlantis* will undertake to link up with *Mir*.

Lemonick, Michael, and Hannah Bloch. "Embrace in Space." *Time*, 10 July 1995. This article reports the docking of Space Shuttle *Atlantis* with the Russian space station *Mir*, orbiting 245 miles (394 kilometers) above Earth, discussing the goals of STS-71 and the political implications of the mission. The article also reviews the history of space cooperation between the United States and Russia since the end of the Cold War.

Linenger, Jerry M. *Off the Planet: Surviving Five Perilous Months Aboard the Space Station Mir*. New York: McGraw-Hill, 2000. In this book, astronaut Jerry M. Linenger describes his background and his experience training for and living aboard *Mir*. Linenger recounts his efforts to survive 132 days aboard the decaying and unstable Russian space station *Mir*.

Morgan, Clay. *Shuttle-Mir: The Illustrated History of the International Space Project*. Book and CD-ROM. NASA SP-2001-4230, Washington, DC, 2001. *http://history.nasa.gov/SP-4225/toc/toc-level1.htm* (accessed 8 March 2012). This book describes Space Shuttle missions to *Mir*, focusing on the people who lived and worked on the Russian space station. Sections of the book are devoted to training, operations, long-duration psychology, and bilingual issues. The book and accompanying CD are full of essays, diagrams, animations, and photographs of *Mir*. The CD contains information about science experiments conducted on the Space Shuttle and on *Mir*; a photo and video collection of Space Shuttle missions and Earth observations; images of mission patches; *High Above Earth*, a children's book; and NASA documents.

Morgan, Clay. *Shuttle-Mir: The United States and Russia Share History's Highest Stage*. Book and CD-ROM. NASA History Series. NASA Report no. SP-2001-4225, Johnson Space Center, NASA, Houston, TX, 2001. *http://history.nasa.gov/SP-4225.pdf* (accessed 11 March 2012). This official NASA history of the 1995–1998 Shuttle-*Mir* program includes mission descriptions and summaries, a description of the support teams in Mission Control, photographs, and interviews. During this period, NASA's Space Shuttles ferried U.S. astronauts to *Mir* so that they could gain experience they would use to build and live on the ISS.

Oberg, James E. "United We Orbit." *Air & Space*, January 1996. The author focuses on Space Shuttle *Atlantis*'s orbital hook-up with the Russian space station *Mir* in June 1995 during STS-71. Topics presented in the article include docking problems, docking mechanisms, design of both countries' probe-drogue systems (devices used to connect two spacecraft to one another), androgynous docking mechanisms, ring-to-ring systems, and remarks made by the crews.

Schuiling, Roelof. "STS-60 Mission Report." *Spaceflight*, April 1994. This article describes STS-60, the first mission of the U.S.-Russian Shuttle-*Mir* program and the first Space Shuttle flight of a Russian cosmonaut, Sergei K. Krikalev. STS-60 carried into orbit the commercially developed Spacehab laboratory module and used it to conduct experiments in the Shuttle payload bay. The author describes crew members, daily activities, launch preparation, and payloads during the nine-day mission.

Schuiling, Roelof. "STS-63 Mission Report." *Spaceflight*, May 1995. This article discusses STS-63, the second mission of the U.S.-Russian Shuttle-*Mir* program. In STS-63, *Discovery* carried out the first rendezvous of a U.S. Space Shuttle with Russia's space station *Mir*. Known as the Near-*Mir* mission, STS-63 carried into orbit Spacehab 3 and the Space Radar Laboratory (SRL). STS-63 was the first flight in which a woman—astronaut Eileen M. Collins—piloted the Shuttle, and the second Shuttle flight carrying a Russian cosmonaut—Vladimir G. Titov. The author describes the nine-day mission's crew members, daily activities, launch preparation, and payloads.

Schuiling, Roelof. "STS-71 Mission Report." *Spaceflight*, October 1995. This article discusses STS-71, the mission of Space Shuttle *Atlantis* to dock with Russia's space station *Mir* and to carry out joint on-orbit operations. This mission was the first time that a U.S. Space Shuttle had docked at *Mir*. The author describes crew members, daily activities, launch preparation, and payloads during the 11-day mission.

Schuiling, Roelof. "STS-74 Mission Report." *Spaceflight*, March 1996. This article discusses STS-74, the mission of Space Shuttle *Atlantis* to dock with Russia's space station *Mir* for the second time and to transfer equipment and supplies to *Mir*. The author describes crew members, daily activities, launch preparation, and payloads during the nine-day mission.

Schuiling, Roelof. "STS-76 Mission Report." *Spaceflight*, July 1996. This article discusses STS-76, the mission of Space Shuttle *Atlantis* to dock with Russia's space station *Mir* for the third time and to deliver the first American female astronaut—Shannon W. Lucid—to live on *Mir*. The author describes crew members, daily activities, launch preparation, and payloads during the 10-day mission.

Schuiling, Roelof. "STS-79 Mission Report." *Spaceflight*, February 1997. This article discusses STS-79, the mission of Space Shuttle *Atlantis* to dock with Russia's space station *Mir* for the fourth time. The Shuttle brought the Spacehab Double Module to support Shuttle-*Mir* activities. The author describes crew members, daily activities, launch preparation, and payloads during the 11-day mission.

Schuiling, Roelof. "STS-81 Mission Report." *Spaceflight*, May 1997. This article discusses STS-81, the mission of Space Shuttle *Atlantis* to dock with Russia's space station *Mir* for the fifth time. The author describes crew members, daily activities, launch preparation, and payloads during the 18-day mission.

Schuiling, Roelof. "STS-84 Mission Report." *Spaceflight*, September 1997. This article discusses STS-84, the mission of Space Shuttle *Atlantis* to dock with Russia's space station *Mir* for the sixth time. The author describes crew members, daily activities, launch preparation, and payloads during the nine-day mission.

Schuiling, Roelof. "STS-86 Mission Report." *Spaceflight*, January 1998. This article discusses STS-86, the mission of Space Shuttle *Atlantis* to dock with Russia's space station *Mir* for the seventh time. The author describes crew members, daily activities, launch preparation, and payloads during the 10-day mission.

Schuiling, Roelof. "STS-89 Mission Report." *Spaceflight*, June 1998. This article discusses STS-89, Space Shuttle *Endeavour*'s mission to dock with the Russian space station *Mir* to deliver scientific equipment, logistical hardware, and water. The author describes crew members, daily activities, launch preparation, and payloads during the eight-day mission.

Schuiling, Roelof. "STS-91 Mission Report." *Spaceflight*, November 1998. This article discusses STS-91, Space Shuttle *Discovery*'s mission to dock with the Russian space station *Mir* and to deliver cargo, science experiments, and supplies. In addition, the crew moved long-term U.S. experiments that had been aboard *Mir* into *Discovery*'s mid-deck locker area. The author describes crew members, daily activities, launch preparation, and payloads during the nine-day mission.

U.S. General Accounting Office. "Former Soviet Union: Information on U.S. Bilateral Program Funding." Report no. GAO/NSIAD-96-37, Washington, DC, 15 December 1995. *http://www.gao.gov/archive/1996/ns96037.pdf* (accessed 8 March 2012). Appendix II of this report lists financial obligations for Space Shuttle support missions with the Russian space program and *Mir*.

"USA, Russia Ink Mir Mission Agreement." *Interavia Business & Technology*, December 1993. This article reviews plans to use U.S. Space Shuttles to supply a new space station, a joint project of the United States and Russia. Merging the U.S. Freedom program and the Russian *Mir-2* program, the two nations have agreed to design and build a space station together. NASA will need to cancel or reschedule other Shuttle missions to fly its Space Shuttles to the planned ISS.

Link to Part 1 (1970–1991), Chapter 13—The Shuttle in International Perspective

CHAPTER 17—THE SPACE SHUTTLE AND THE INTERNATIONAL SPACE STATION

"A Black Hole in the Sky." *Economist*, 14 November 1998. NASA is upgrading the Space Shuttle fleet to enable the Shuttles to carry parts needed to build the ISS, thereby reducing the space station's dependence on unpiloted Russian supply ships for refueling. NASA may delay essential repairs to the HST so that it can send additional Shuttle flights to help build the ISS.

Achenbach, Joel. "Retirement of Shuttles a Risk for Space Station, Critics Say." *Washington Post*, 4 July 2011. This article reports that Christopher C. Kraft Jr., former Director of NASA's Johnson Space Center, has cowritten a letter, endorsed by a number of NASA veterans and astronauts of the *Apollo* era, contending that the ISS will become more hazardous for astronauts without the availability of Space Shuttle resources for emergency backup.

"Another Node for the Space Station." *Interavia Business & Technology*, Winter 2007. The article reports on the installation of the ISS's Harmony Node 2, the interconnecting unit developed in Italy by Thales Alenia Space. Node 2, a crucial element for the development and completion of the orbital structure, will provide a passageway connecting the three science laboratories: the United States' Destiny, Europe's Columbus, and Japan's Kibo modules.

Anselmo, Joseph C. "NASA Confident of Shuttle Backups." *Aviation Week & Space Technology*, 8 April 1996. NASA has scheduled 26 Space Shuttle flights over the next four years, to enable assembly of the ISS. However, some observers are concerned that unanticipated problems could interfere with the Shuttle launch schedule, which is essential for the construction of the space station. Many key ISS components, such as the U.S. laboratory and habitation modules and connecting nodes, are designed specifically for the Shuttle's payload bay, so grounding the Shuttle for any significant amount of time would delay space station assembly and provide ammunition to the ISS's opponents in Congress.

Asker, James R. "Space Station Key to NASA's Future." *Aviation Week & Space Technology*, 15 March 1993. The author discusses NASA's budgetary concerns, suggesting that NASA will have to make difficult choices about program priorities. He is concerned that, if NASA dedicates its budget to building a space station, it will have inadequate funds for other programs, including aeronautics, astronomy, and exploration. However, NASA Administrator Daniel S. Goldin envisions NASA launching six major scientific spacecraft and a new generation of smaller ones over several years, promoting interest in space sciences and engineering among graduate students, and fostering new enthusiasm for science and mathematics among school children.

Behrens, Carl E. "The International Space Station and the Space Shuttle." CRS Report for Congress, Congressional Research Service, Library of Congress, Washington, DC, 18 March 2009. *http://www.fas.org/sgp/crs/space/RL33568.pdf* (accessed 8 March 2012). This report details the developments, designs, costs, and schedules of the ISS, as well as describing the future of the Space Shuttle and its future budgets. To conclude, the report outlines Space Shuttle issues for Congress to consider.

Behrens, Carl, and Mary Beth Nitikin. "Extending NASA's Exemption from the Iran, North Korea, and Syria Nonproliferation Act." CRS Report for Congress, Congressional Research Service, Library of Congress, Washington, DC, 1 October 2008. *http://assets. opencrs.com/rpts/RL34477_20081001.pdf* (accessed 8 March 2012). This report discusses the Iran Nonproliferation Act of 2000, enacted to prevent foreign transfers to Iran of weapons of mass destruction, missile technology, and advanced conventional weapons technology, particularly transfers from Russia. Section 6 of the Act bans U.S. payments to Russia in connection with the ISS unless the President of the United States determines that Russia is taking steps to prevent the proliferation of weapons to Iran.

Covault, Craig. "Complex ISS Assembly Flight Poised for Liftoff." *Aviation Week & Space Technology*, 30 November 1998. The author claims that STS-80, the first Space Shuttle mission to assemble large elements of the ISS, will require precision flying, extensive robotics, and orbital construction, to join and outfit the station's initial U.S. and Russian modules. Shuttle astronauts will use the Canadian arm to lift the 15-by-33-foot (4.6-by-10-meter), 12.5-ton (11.3-tonne or 11,340-kilogram) Unity module out of the Shuttle's aft payload bay and affix it to the Russian Zarya spacecraft.

Covault, Craig. "ISS Assembly Readied as Shuttle Pace Is Assessed." *Aviation Week & Space Technology*, 7 December 1998. The article discusses the ISS project, the largest international aerospace project ever undertaken, involving 16 nations and nearly 300 prime and subcontractors from around the world. The author argues that NASA will have difficulty meeting the schedule of three dozen Space Shuttle flights within the tight rendezvous launch windows required for ISS operations.

Covault, Craig. "Station Commercialization Set as Assembly Flight Readied." *Aviation Week & Space Technology*, 30 November 1998. This article describes *Endeavour*'s STS-88, the first Space Shuttle mission to assemble the initial elements of the ISS and discusses NASA's plan to turn the multibillion-dollar ISS into a commercial, fee-for-service facility. According to this scenario, private industry would assume substantial responsibility for the space station.

Covault, Craig. "Station Training Focus: 'Building on the Fly.'" *Aviation Week & Space Technology*, 2 September 1996. This article describes astronaut and cosmonaut training for those who will build and live in the ISS. The author presents information about the Space Shuttle Training Facility at NASA's Johnson Space Center in Houston, Texas, and the Neutral Buoyancy Laboratory underwater training facility. He also explains EVA training requirements. In addition, the article discusses the costs of the ISS and its assembly schedule.

Evans, Ben. "STS-92: The Builders Move into the ISS." *Spaceflight*, September 2000. This article describes STS-92, a mission in which Space Shuttle *Discovery* carried two new modules to add to the ISS. *Discovery*'s crew installed the Z-1 truss, the first of 10 prefabricated sections of the gigantic Integrated Truss Segment.

Geenty, John. "Flights of Fancy." *Spaceflight*, January 2005. This article describes how NASA planned its Space Shuttle missions before the *Columbia* tragedy. Originally, NASA had planned for the Shuttle to fly more often, with a greater range of payloads. However, the losses of *Challenger* and *Columbia* led to a change in Shuttle flights and schedules. Moreover, the loss of *Columbia* meant that future Shuttle missions would have to support the ISS, with no flights to spare for science programs.

Hansen, Rosanna. *Liftoff! A Space Adventure*. Pleasantville, NY: Reader's Digest Children's Books, 2000. Ages 7–8. This book invites young readers to imagine blasting off into space aboard a Space Shuttle on a flight to help build a space station. The book describes experiencing G-forces when lifting off into space, how it feels to float in microgravity while orbiting the Earth in a Shuttle, suiting up and going outside of the Shuttle to help build the space station, moving around with the help of a jetpack, eating a meal in space, and napping in a sleeping bag strapped to the wall of the crew quarters to prevent floating around.

Jones, Thomas D. "The Future of NASA's Astronaut Corps." *Aerospace America*, October 2010. In this article, a former astronaut compares the number of astronauts at NASA when he flew his first Shuttle flight in 1994 with the number of astronauts that NASA will need after the Space Shuttles retire. Jones also discusses astronaut hiring and training for ISS missions.

Kidger, Neville. "STS-119: Orbital Operations." *Spaceflight*, May 2009. This article describes activities aboard the ISS, as well as summarizing STS-119, Space Shuttle *Discovery*'s mission to deliver solar arrays to the ISS. The article lists the Shuttle crew members and describes the crew's activities during the 12-day mission.

Kidger, Neville. "STS-120: Orbital Operations." *Spaceflight*, January 2008. This article describes activities aboard the ISS, as well as summarizing STS-120, Space Shuttle *Discovery*'s mission to deliver Harmony Node, a connecting module that will increase the orbiting laboratory's interior space. STS-120 crew also repaired the ISS's solar arrays. The article lists the STS-120 crew members and describes day-to-day operations during the 16-day mission.

Kidger, Neville. "STS-122: Orbital Operations." *Spaceflight*, April 2008. This article describes activities aboard the ISS, as well as summarizing STS-122, Space Shuttle *Atlantis*'s mission to deliver and to install the Columbus laboratory, the ISS module contributed by ESA. The article lists the STS-122 crew members and describes the crew's activities during the 13-day mission.

Kidger, Neville. "STS-123: Orbital Operations." *Spaceflight*, May 2008. This article describes activities aboard the ISS, as well as summarizing STS-123, Space Shuttle *Endeavour*'s mission to equip and supply the ISS. *Endeavour* delivered the Japanese Experiment

Logistics Module, which contains avionics and will serve as a storage area for experiment materials. The article, which also names STS-123 crew members and describes day-to-day operations during the 15-day mission, continues in the June 2008 edition of *Spaceflight*.

Kidger, Neville. "STS-123: Orbital Operations." *Spaceflight*, June 2008. This article continues an article in the May 2008 edition of *Spaceflight* summarizing STS-123, Space Shuttle *Endeavour*'s mission to equip and supply the ISS. *Endeavour* delivered the Japanese Experiment Logistics Module, which contains avionics and will serve as a storage area for experiment materials. The article lists STS-123 crew members and describes day-to-day operations during the 15-day mission.

Kidger, Neville. "STS-127: Orbital Operations." *Spaceflight*, October 2009. This article describes activities aboard the ISS, as well as summarizing STS-127, Space Shuttle *Endeavour*'s mission to deliver the Kibo Japanese Experiment Module Exposed Facility and Experiment Logistics Module Exposed Section. The article lists the Shuttle crew members and describes the crew's activities during the 15-day mission.

Kidger, Neville. "STS-128: Orbital Operations." *Spaceflight*, November 2009. This article describes activities aboard the ISS, as well as summarizing STS-128, Space Shuttle *Discovery*'s mission to deliver 7 tons (6.4 tonnes) of equipment and supplies to the ISS. After this mission, NASA will begin a transition from Shuttle missions to help build the ISS, to missions to use the ISS. The article lists crew members, recounts prelaunch activities, and describes activities during the 13-day mission.

Klotz, Irene. "Crunch Time for Shuttle as Safety Demands Pile Up." *New Scientist*, 15 July 2006. The author describes NASA's reactions to Space Shuttle *Discovery*'s STS-114 return-to-flight mission. Instead of experiencing jubilation, NASA personnel are bracing for the effort they must make to complete the ISS before the Shuttle's retirement in 2010.

Kremer, Ken. "STS-129: Shuttle Delivers Spares to ISS." *Spaceflight*, February 2010. This article describes STS-129, the mission of Space Shuttle *Atlantis* to equip and supply the ISS. The article lists crew members and daily activities during the 10-day mission.

Kremer, Ken. "STS-130: New Window on the World." *Spaceflight*, April 2010. This article describes STS-130, Space Shuttle *Endeavour*'s mission to deliver equipment and supplies to the ISS. The article describes prelaunch activities, crew members, and activities during the 13-day mission.

Kremer, Ken. "STS-131: Discovery's Penultimate Voyage." *Spaceflight*, June 2010. This article describes STS-131, in which Space Shuttle *Discovery* delivered 8 tons (7.3 tonnes) of equipment and supplies to the ISS. The article describes prelaunch activities, crew members, and day-to-day operations during the 15-day mission.

Kremer, Ken. "STS-132: Atlantis' Last Blast with Russian Beauty." *Spaceflight*, August 2010. This article describes STS-132, Space Shuttle *Atlantis*'s final mission before its retirement. The goal of STS-132 was to equip and supply the ISS. Besides carrying spare parts to the ISS, *Atlantis* delivered the 11,000-pound (4.989.5-kilogram) Russian Rassvet Mini Research Module, which will enable ISS crew to conduct biotechnology and fluid physics experiments. The article describes prelaunch activities, crew members, payloads, and daily operations during the 11-day mission.

Kremer, Ken, and Gerard van de Haar. "STS-133: Discovery's Final Voyage into Orbit." *Spaceflight*, May 2011. This article describes STS-133, Space Shuttle *Discovery*'s mission to deliver spare parts to the ISS. The article describes prelaunch activities, crew members, and activities during the 12-day mission. Space Shuttle *Discovery* retired after completing this mission.

Lawler, Andrew. "NASA May Cut Shuttle Flights and Reduce Science on Station." *Science* 309, no. 5734 (22 July 2005): 540–541. This article focuses on NASA's plan to curtail Space Shuttle flights and to reduce science research activities on the ISS.

Lawler, Andrew, Daniel Clery, and Dennis Normile. "Life Science Research on Space Station Is Headed for Big Cuts." *Science* 308, no. 5722 (29 April 2005): 610–611. The authors report that NASA is finalizing a new plan to reduce the quality and quantity of cutting-edge research on the ISS. The cost of returning the Space Shuttle to flight, finishing the ISS by 2010, and building new launchers will have a negative effect on continuing high-priority research in biology at the space station.

Leary, Warren E. "Shuttle To Put European Module in Space." *New York Times*, 5 December 2007. Space Shuttle *Atlantis* is poised for a 6 December 2007 launch to the ISS on an 11-day mission that will make the orbital outpost more international. The principal goal of the mission, STS-117, is to add ESA's new Columbus laboratory to the growing station.

Logsdon, John M. "Lessons To Be Learned from Space Station Saga." *Aviation Week & Space Technology*, 7 March 1994. The author claims that, in 1983, when President Ronald W. Reagan ordered NASA to build a space station, European nations also declared their intention to build a space station. Yet, 10 years later, neither the United States nor Europe has followed through. The author suggests that the lack of powerful supporters for the project and disagreement about the value of building a space station have thwarted these plans.

McElroy, John H. "Some Thoughts on Space Station Science." *Space Policy* 17, no. 4 (November 2001): 257–260. The author comments that ISS budget cuts negatively affect U.S. scientific capabilities and undermine the United States' investment in the Space Shuttle.

Morring, Frank, Jr. "Commercial Break; NASA Plans COTS-Only Approach for ISS, Dropping Russia's Progress." *Aviation Week & Space Technology*, 21 April 2008. This article reports that NASA officials will discuss with Congress a plan for NASA to continue using Russia's Soyuz crew launch vehicles to transport astronauts to and from the ISS after the final Space Shuttle flight in 2010. NASA does not intend to continue using

Russian Progress vehicles for U.S. cargo resupply but plans to use its own Commercial Orbital Transportation System (COTS) program vehicles, which are as yet untested.

Morring, Frank, Jr. "End Game; Space Station Managers Tread Tricky Path to Completion with 10 Shuttle Flights Left." *Aviation Week & Space Technology*, 31 March 2008. This article describes the remaining scheduled Space Shuttle missions to build the ISS. The Shuttles will deliver spare parts so that they will be available on the ISS when original hardware wears out. The author notes that NASA is still having difficulty incorporating the required safety modifications to the foam-covered external tanks that carry propellants for Shuttle launches.

Morring, Frank, Jr. "Finishing the Job; Long-Planned Station-Assembly Finale Promises To Be a Nail-Biter for Crews." *Aviation Week & Space Technology*, 21 August 2006. The author of this article reports that NASA astronauts, engineers, and controllers intend to try to double the size of the ISS by adding as many as five pressurized modules and three more large solar arrays to power them. Transporting and installing the equipment will require more than 16 flights over four years and the labor of crews from Canada, Europe, Japan, Russia, and the United States. The Space Shuttle is the only spacecraft able to transport the large components of the ISS, which were custom-built for its cargo bay.

Morring, Frank, Jr. "One More Time; Hubble Program Planning Final Shuttle Mission as Early as 2007 To Maintain the Observatory." *Aviation Week & Space Technology*, 11 July 2005. This article describes NASA's decision to approve a Shuttle mission to repair the HST. The author also discusses NASA's efforts to schedule a realistic number of Shuttle flights to enable the ISS partners to complete assembly of the space station before the Shuttle's planned retirement at the end of 2010.

Morring, Frank, Jr. "Shuttle Accident Puts 'Everything on Table'." *Aviation Week & Space Technology*, 10 February 2003. This article describes the efforts of Congress and the presidential administration to determine the cause of the *Columbia* accident. The congressional inquiry covered technical issues and political issues underlying space policy decisions on matters such as construction of the ISS, the need for RLVs, and funding to ensure Space Shuttle safety. The accident halted NASA's policy reviews regarding the final configuration of the ISS and plans to build an Orbital Space Plane.

Mowbray, Scott. "After Columbia: The ISS in Crisis." *Popular Science*, April 2003. This article focuses on how Space Shuttle *Columbia*'s accident has affected the construction of the ISS. The author explains that NASA had intended 2003—the year of the *Columbia* tragedy—as a pivotal year for building the space station. Instead, NASA will have to work with Russia's space authorities to ensure the safe operations of the ISS during the long grounding before another Shuttle is cleared to return to space.

"New ISS Plan Based on 16 Shuttle Flights." *Interavia Business & Technology*, Spring 2006. This article reports that the heads of the national space agencies collaborating on the ISS have announced the planned sequence of spaceflights to assemble the new space station. The agencies intend to use the space transportation systems of Europe, Japan, Russia, and the United States to ensure full use of the ISS. Besides using U.S. and Russian spacecraft to transport cargo to the ISS, the partners will use the European automated cargo vehicle, which is scheduled for launch on an Ariane rocket in the spring of 2007. The U.S. Space Shuttle will make a series of 16 flights to carry Japan's experiment module Kibo and Europe's Columbus laboratory to the space station.

Reichhardt, Tony. "Columbia Explosion May Trigger Fatal Delays for Space Station." *Nature* 421, no. 6923 (6 February 2003): 561. This article focuses on the effect of Space Shuttle *Columbia*'s accident on the ISS program and on NASA's other projects.

Reichhardt, Tony. "Researchers Find Silver Lining in Delay To Work on Space Station." *Nature* 386, no. 6626 (17 April 1997): 633. This article describes the reaction to NASA's decision to delay the construction of the ISS. The good news for NASA is that the announced 11-month delay will give NASA time to conduct additional Space Shuttle research missions.

Reichhardt, Tony. "Shuttlenauts: The Faces of the Space Shuttle Era." *Air & Space*, January 2011. The author discusses the Shuttle's role in the construction of the ISS and examines the careers of several astronauts who spent time on the Space Shuttle, including former Navy pilot Robert L. Crippen and Shuttle Commander Peggy A. Whitson.

Reuters. "Panel Agrees To Protect Jobs, Add Mission for Space Shuttle." *Washington Post*, 16 July 2010. Reuters reports that the Senate Commerce, Science, and Transportation Committee has unanimously passed a plan to postpone retirement of the Space Shuttle. In addition, the committee has agreed to include in NASA's three-year spending plan an additional Space Shuttle mission to the ISS.

Schuiling, Roelof. "STS-88: 'Unity' Module Delivered to Space Station." *Spaceflight*, March 1999. This article discusses STS-88, Space Shuttle *Endeavour*'s mission to begin construction of the ISS. The author describes crew members, daily activities, launch preparation, and payloads during the 13-day mission.

Schuiling, Roelof. "STS-92: Discovery Completes 100th Shuttle Flight." *Spaceflight*, January 2001. This article describes STS-92, Space Shuttle *Discovery*'s mission to continue the construction of the ISS. The article describes launch preparation, payloads, crew members, and daily activities during the 14-day mission.

Schuiling, Roelof. "STS-96: Discovery Shuttle Launched on First Mission of 1999." *Spaceflight*, October 1999. This article discusses STS-96, Space Shuttle *Discovery*'s mission to deliver parts and materials to the ISS. The author describes crew members, day-to-day operations, launch preparation, and payloads during the nine-day mission.

Schuiling, Roelof. "STS-97: Endeavour Delivers Solar Arrays to ISS." *Spaceflight*, March 2001. This article discusses STS-97, Space Shuttle *Endeavour*'s mission to deliver solar arrays and supplies to the ISS. The author describes crew members, daily activities, launch preparation, and payloads during the 12-day mission.

Schuiling, Roelof. "STS-98: Atlantis Carries Destiny to Orbit." *Spaceflight*, May 2001. This article describes STS-98, the mission of Space Shuttle *Atlantis* to take the United States' laboratory Destiny to the ISS. The article names the crew and describes their activities for each of the 14 days the Shuttle was in space.

Schuiling, Roelof. "STS-100: Endeavour Carries Robotic Components to ISS." *Spaceflight*, August 2001. This article discusses STS-100, Space Shuttle *Endeavour*'s mission to equip and supply the ISS. The author describes crew members, daily activities, launch preparation, and payloads during the 13-day mission.

Schuiling, Roelof. "STS-101: Maintenance Mission to Unity-Zarya." *Spaceflight*, August 2000. This article describes STS-101, the mission of Space Shuttle *Atlantis* to maintain the Unity-Zarya ISS. Unity and Zarya were the first ISS components. The article describes launch preparation, payloads, crew members, and activities during the 11-day mission.

Schuiling, Roelof. "STS-104: Atlantis Delivers Space Doorway to ISS." *Spaceflight*, October 2001. This article describes STS-104, the mission of Space Shuttle *Atlantis* to deliver and install a space doorway on the ISS. The article describes launch preparation, payloads, crew members, and activities during the 14-day mission.

Schuiling, Roelof. "STS-105: Space Shuttle Mission Report." *Spaceflight*, November 2001. This article describes STS-105, Space Shuttle *Discovery*'s mission to equip and supply the ISS. The article describes launch preparation, payloads, crew members, and daily activities during the 13-day mission.

Schuiling, Roelof. "STS-106: Atlantis Revisits Space Station." *Spaceflight*, December 2000. This article describes STS-106, the mission of Space Shuttle *Atlantis* to equip the ISS and prepare the station for its first crew. The article describes launch preparation, payloads, crew members, and activities during the 13-day mission.

Schuiling, Roelof. "STS-108: Endeavour Carries Fourth Crew to Space Station." *Spaceflight*, March 2002. This article discusses STS-108, Space Shuttle *Endeavour*'s mission to equip and supply the ISS. The author describes crew members, daily activities, launch preparation, and payloads during the 13-day mission.

Schuiling, Roelof. "STS-111: Weather Delays Endeavour's Launch and Landing." *Spaceflight*, September 2002. This article discusses STS-111, Space Shuttle *Endeavour*'s mission to equip and supply the ISS. The author describes crew members, daily activities, launch preparation, and payloads during the 15-day mission.

Schuiling, Roelof. "STS-112: Advances Space Station Assembly." *Spaceflight*, January 2003. This article describes STS-112, the mission of Space Shuttle *Atlantis* to deliver to the ISS the S1 integrated truss segment and spacewalk platform and to install the new components. The article describes launch preparation, payloads, crew members, and activities during the 12-day mission.

Schuiling, Roelof. "STS-113: Endeavour Overcomes Technical Hitches To Deliver Sixth ISS Crew." *Spaceflight*, March 2003. This article discusses STS-113, Space Shuttle *Endeavor*'s mission to equip and supply the ISS. The author describes crew members, daily activities, launch preparation, and payloads during the 15-day mission.

Shayler, David J. *Walking in Space*. New York: Springer-Praxis, 2004. This book provides a comprehensive overview and analysis of the techniques astronauts use in EVAs (spacewalks). The author draws on original documentation and on personal interviews with astronauts who have EVA experience, as well as on the accounts of staff involved in spacesuit design, EVA planning, and operations. The book describes the development of techniques for ensuring crew safety during spacewalks and looks ahead to future EVAs from the ISS and the development of new technology.

Sietzen, Frank, Jr. "The Future of Space Transportation: Is It Expendable? Reusables? Or the Shuttle? Why Not All Three?" *Ad Astra*, August 2002. This short article discusses the future of the Space Shuttle and Congress's failure to increase NASA's budget, despite rising ISS costs.

Simpson, Clive. "STS-114: Return of the Space Shuttle." *Spaceflight*, October 2005. This article discusses STS-114, Space Shuttle *Discovery*'s mission to deliver equipment and supplies to the ISS. The author describes crew members, daily activities, launch preparation, and payloads during the 13-day mission.

Simpson, Clive. "STS-115: Astronauts Complete Tough Mission." *Spaceflight*, November 2006. This article discusses STS-115, the mission of Space Shuttle *Atlantis* to deliver to the ISS the P3/P4 integrated truss and to install the truss and a pair of solar arrays on the ISS. The author describes crew members, daily activities, launch preparation, and payloads during the 11-day mission.

Simpson, Clive. "STS-117: Atlantis Completes Spectacular Mission." *Spaceflight*, August 2007. This article discusses STS-117, the mission of Space Shuttle *Atlantis* to deliver to the ISS and to install the second and third starboard truss segments S3/S4 and another pair of solar arrays. The author describes crew members, daily activities, launch preparation, and payloads during the 17-day mission.

Simpson, Clive, and Gerard van der Haar. "STS-122: Columbus Ushers in a New Space Era." *Spaceflight*, April 2008. The authors of this article describe STS-122, Space Shuttle *Atlantis*'s mission to carry the Columbus laboratory to the ISS. The crew of the ISS took three spacewalks to prepare the 21-ton (19.05-tonne) Columbus laboratory for scientific work. The authors describe crew activities over the 13 days the Shuttle was in space.

Simpson, Clive, and Tim Furniss. "STS-121: Harmony Launch Starts Busy Period for Europe." *Spaceflight*, December 2007. This article describes *Discovery*'s STS-121 mission to carry to the ISS the Harmony node, a passageway connecting the three ISS science laboratories. The article describes crew activities over the 14 days that the Shuttle was in space.

Simpson, Clive, Gerard van der Haar, and Rudolf van Beest. "STS-116: Shuttle Mission Re-Wires ISS." *Spaceflight*, February 2007. This article discusses STS-116, Space Shuttle *Discovery*'s mission to deliver equipment and supplies to the ISS. The author describes crew members, daily activities, launch preparation, and payloads during the 12-day mission.

Simpson, Clive, Gerard van der Haar, and Rudolf van Beest. "STS-118: Endeavour Finally Returns to Space." *Spaceflight*, October 2007. This article discusses STS-118, Space Shuttle *Endeavour*'s mission to deliver equipment and supplies to the ISS. The crew delivered and installed a third starboard truss segment and 5,000 pounds (2,268 kilograms) of equipment and supplies. The author describes crew members, daily activities, launch preparation, and payloads during the 12-day mission.

Smith, Marcia S. "Space Stations." CRS Issue Brief for Congress, Congressional Research Service, Library of Congress, Washington, DC, 4 January 2006. *http://assets.opencrs. com/rpts/IB93017_20060104.pdf* (accessed 7 March 2012). This report discusses developments, designs, costs, and schedules of the ISS, as well as the future of the Space Shuttle and its future budgets. The author also reports related congressional actions of FY 2005 and FY 2006. In addition, the report discusses the space station's international partners, including Canada, Europe, and Japan, and examines the risks and benefits of partnering with Russia.

Smith, Marcia S., Daniel Morgan, and Wendy H. Schacht. "The National Aeronautics and Space Administration's FY 2004 Budget Request: Description, Analysis, and Issues for Congress." CRS Report for Congress, Congressional Research Service, Library of Congress, Washington, DC, 23 September 2003. *http://assets.opencrs.com/rpts/ RL31821_20030923.pdf* (accessed 8 March 2012). This report discusses NASA's budget request of US$15.469 billion for FY 2004. NASA is making this budget request against the backdrop of the Space Shuttle *Columbia* tragedy, a context that could significantly influence NASA's appropriation. Other NASA budget issues include funding for the ISS program, Project Prometheus, aeronautics, and technology transfer.

U.S. Congress. Congressional Budget Office. "A Budgetary Analysis of NASA's New Vision for Space Exploration." Report, Washington, DC, 2 September 2004. *http://www.cbo.gov/ ftpdocs/57xx/doc5772/09-02-NASA.pdf* (accessed 8 March 2012). This report provides an analysis of NASA's budget request for FY 2005, as well as an analysis of NASA's budget projection, which forecasts budgetary requirements through 2020. In its analysis, the CBO assesses the implications of NASA's budget plans for the content and schedule of NASA's future activities, including the operation of the Space Shuttle and the United States' participation in the ISS.

U.S. Congress. House of Representatives. Committee on Science. *NASA's Fiscal Year 2007 Budget Proposal.* 109th Cong., 2nd sess., 16 February 2006. *http://www.gpo.gov/fdsys/pkg/ CHRG-109hhrg25937/pdf/CHRG-109hhrg25937.pdf* (accessed 12 March 2012). This hearing presents an overview of NASA's FY 2007 budget request, covering issues related to NASA programs, including the status of the ISS, the Space Shuttle, and crew exploration vehicle programs.

U.S. Congress. House of Representatives. Committee on Science. *Status of NASA's Programs.*
109th Cong., 1st sess., 3 November 2005. *http://www.gpo.gov/fdsys/pkg/CHRG-109hhrg24151/pdf/CHRG-109hhrg24151.pdf* (accessed 12 March 2012). This hearing
provides an overview of NASA's approach to implementing the President's New Vision
for Space Exploration. The hearing reviews issues related to NASA programs, including
the status of the ISS, the Space Shuttle, and crew exploration vehicle programs.

U.S. Congress. House of Representatives. Committee on Science. *The Status of Russian
Participation in the International Space Station Program.* 105th Cong., 1st sess., 2
February 1997. *http://www.gpo.gov/fdsys/pkg/CHRG-105hhrg38882/pdf/CHRG-105hhrg38882.pdf* (accessed 10 April 2012). This hearing focuses on congressional
oversight of the ISS, Russia's difficulties meeting its commitments to the ISS, and the
effect of Russia's difficulties on U.S. efforts to support the space station.

U.S. Congress. House of Representatives. Committee on Science. *U.S.-Japanese Cooperation in
Human Spaceflight.* 104th Cong., 1st sess., 19 October 1995. *https://web.lexis-nexis.com/
congcomp/* (accessed 11 November 2011 via Proquest Congressional). This hearing
provides background on U.S.-Japanese cooperative space programs, including the status
of these programs and the outlook for their future. Specifics of Japanese participation in
ISS scientific experiments are also included.

U.S. Congress. House of Representatives. Committee on Science. *U.S.-Russian Cooperation in
Human Spaceflight.* 105th Cong., 1st sess., 18 September 1997 and 6 May, 24 June, 5
August, and 7 October 1998. *http://catalog.gpo.gov/fdlpdir/locate.jsp?ItemNumber=
1025-A-01&SYS=000502274* (accessed 11 November 2011). These hearings review the
status of NASA's ISS program, focusing on problems related to Russian Government
participation in the ISS program, including the inability to meet funding obligations.

U.S. Congress. House of Representatives. Committee on Science and Technology. *NASA's
Fiscal Year 2008 Budget Request.* 110th Cong., 1st sess., 15 March 2007. *http://frwebgate.
access.gpo.gov/cgi-bin/getdoc.cgi?dbname=110_house_hearings&docid=f:33803.pdf*
(accessed 13 March 2012). This hearing, an overview of NASA's FY 2008 budget
request, focuses on issues related to NASA programs, including the status of the ISS, the
Space Shuttle, and crew exploration vehicle programs. The hearing also discusses
concerns about potential funding shortfalls in NASA's FY 2008 budget request and the
effect of these shortfalls on NASA programs.

U.S. Congress. House of Representatives. Committee on Science and Technology. *NASA's
Fiscal Year 2009 Budget Request.* 110th Cong., 2nd sess., 13 February 2008. *http://www.
gpo.gov/fdsys/pkg/CHRG-110hhrg40598/html/CHRG-110hhrg40598.htm* (accessed 8
March 2012). This hearing, an overview of NASA's FY 2009 budget request, covers
issues related to NASA programs, including the status of the ISS, the Space Shuttle, and
exploration and research programs. The request includes US$2.98 billion to operate and
maintain NASA's three Space Shuttles.

U.S. Congress. House of Representatives. Committee on Science and Technology. *NASA's Fiscal Year 2010 Budget Request*. 111th Cong., 1st sess., 19 May 2009. *http://www.gpo. gov/fdsys/pkg/CHRG-111hhrg49551/pdf/CHRG-111hhrg49551.pdf* (accessed 13 March 2012). This hearing, an overview of NASA's FY 2010 budget request, covers issues related to NASA programs, including the status of the ISS, the Space Shuttle, and exploration and research programs.

U.S. Congress. House of Representatives. Committee on Science and Technology. *NASA's Fiscal Year 2011 Budget Request and Issues*. 111th Cong., 2nd sess., 25 February 2010. *http://frwebgate.access.gpo.gov/cgi-bin/getdoc.cgi?dbname=111_house_hearings& docid=f:55837.pdf* (accessed 8 March 2012). This hearing examines NASA's FY 2011 budget request, covering issues related to NASA programs, including the status of the ISS, the Space Shuttle, and exploration and research programs.

U.S. Congress. House of Representatives. Committee on Science and Technology. Subcommittee on Space and Aeronautics. *NASA's International Space Station Program: Status and Issues*. 110th Cong., 2nd sess., 24 April 2008. *http://www.gpo.gov/fdsys/pkg/ CHRG-110hhrg41799/pdf/CHRG-110hhrg41799.pdf* (accessed 7 March 2012). This hearing examines the challenges and risks facing the ISS program in light of the decision to retire the Space Shuttle in 2010.

U.S. Congress. House of Representatives. Committee on Science and Technology. Subcommittee on Space and Aeronautics. *NASA's Space Shuttle and International Space Station Programs: Status and Issues*. 110th Cong., 1st sess., 24 July 2007. *http://www.gpo. gov/fdsys/pkg/CHRG-110hhrg36737/html/CHRG-110hhrg36737.htm* (accessed 14 March 2012). This hearing examines the main challenges to NASA's accomplishing its major goals—to continue successfully flying the Space Shuttle until its planned retirement in 2010 and to complete the planned Shuttle mission to service the HST. This hearing also considers the obstacles to completing the assembly of the ISS by the time NASA retires the Space Shuttle.

U.S. Congress. House of Representatives. Committee on Science, Space, and Technology. Subcommittee on Space. *1993 NASA Authorization, Volume II*. 102nd Cong., 2nd sess., 19 February–5 March 1992. *https://web.lexis-nexis.com/congcomp/* (accessed 11 November 2011 via Proquest Congressional). These hearings provide highlights of the 22 January 1992 Space Shuttle mission, STS-42, a mission focusing on scientific experiments. They also review the status and accomplishments of the ISS program and provide an overview of Space Shuttle programs, focusing on cost-reduction issues and the proposed cancellation of the advanced solid rocket motor program.

U.S. Congress. Senate. Committee on Commerce, Science, and Transportation. Subcommittee on Science and Space. *Assessing Commercial Space Capabilities*. 111th Cong., 2nd sess., 18 March 2010. *http://frwebgate.access.gpo.gov/cgi-bin/getdoc.cgi?dbname=111_senate_ hearings&docid=f:66983.pdf* (accessed 14 March 2012). This hearing examines commercial space capabilities and developments in light of the President George W. Bush administration's decision to discontinue the Space Shuttle program in favor of expanded commercial-sector contracting, under NASA oversight, for transportation of astronauts to the ISS and for further exploration.

U.S. Congress. Senate. Committee on Commerce, Science, and Transportation. Subcommittee on Space, Aeronautics, and Related Sciences. *Issues Facing the U.S. Space Program After Retirement of the Space Shuttle*. 110th Cong., 1st sess., 15 November 2007. In this hearing, NASA witnesses discuss various aspects of the U.S. space program after NASA retires the Space Shuttle, including the status of space transportation in support of the ISS.

U.S. General Accounting Office. "International Space Station: U.S. Life-Cycle Funding Requirements." Report no. GAO/NSIAD-98-147, Washington, DC, May 1998. *http://www.gao.gov/archive/1998/ns98147.pdf* (accessed 7 March 2012). This report on the ISS includes information about the cost of Shuttle flights and their effect on the cost of the space station. The GAO finds that, without the offsetting reduction in Shuttle support costs, the overall costs of the space station would have been significantly higher.

U.S. General Accounting Office. "Major Management Challenges and Program Risks: National Aeronautics and Space Administration." Report no. GAO-01-258, Washington, DC, January 2001. *http://www.gao.gov/pas/2001/d01258.pdf* (accessed 8 March 2012). This report indicates that, since 1995, the Shuttle workforce has decreased by more than one-third. Several internal NASA studies have shown that the reduction has negatively affected the Space Shuttle program's workforce. Many key program areas have insufficient qualified staff, and the remaining staff shows signs of overwork and fatigue. Moreover, the skill mix and demographics of the Shuttle workforce jeopardize NASA's ability to increase the Shuttle flight rate in support of the ISS's assembly and to transfer leadership roles to the next generation.

U.S. General Accounting Office. "NASA: Shuttle Fleet's Safe Return to Flight Is Key to Space Station Progress." Report no. GAO-04-201T, Washington, DC, 29 October 2003. *http://www.gao.gov/new.items/d04201t.pdf* (accessed 7 March 2012). This report addresses concern over the cost and assembly schedule for the ISS after NASA grounded Space Shuttle flights following the *Columbia* accident. The report also discusses the implications of the Shuttle fleet's grounding on the schedule and cost of building the ISS and on the ISS partners' funding and agreements.

U.S. General Accounting Office. "Space Shuttle: Declining Budget and Tight Schedule Could Jeopardize Space Station Support." Report no. GAO/NSIAD-95-171, Washington, DC, 28 July 1995. *http://www.gao.gov/archive/1995/ns95171.pdf* (accessed 8 March 2012). This report reviews NASA's efforts to redesign the lift capability of the Space Shuttle so that it will be able to make the 21 flights necessary to complete the assembly of the ISS within five years. The GAO found that NASA's plans for increasing the Shuttle's lift capability are complex, involving approximately 30 individual actions, such as hardware redesigns, improved flight-design techniques, and new operational procedures.

U.S. General Accounting Office. "Space Shuttle: NASA Must Reduce Costs Further To Operate Within Future Projected Funds." Report no. GAO/NSIAD-95-118, Washington, DC, June

1995. *http://archive.gao.gov/t2pbat1/154853.pdf* (accessed 8 March 2012). This report assesses Space Shuttle program cost reductions, past and future, and the effects of these reductions on safety. So far, NASA has reduced its cumulative funding for Shuttle operations by 22 percent, from FY 1992 to FY 1995, and its actual operating costs by 8.5 percent, between FY 1992 and 1994. The GAO report found that, although additional funding reductions are necessary to achieve NASA's future budget projections, NASA may not be able to reduce its costs without affecting Shuttle safety and jeopardizing the ISS program.

U.S. General Accounting Office. "Space Shuttle: Upgrade Activities and Carryover Balances." Report no. GAO/T-NSIAD-98-21, Washington, DC, 1 October 1997. *http://www.gao. gov/archive/1998/ns98021t.pdf* (accessed 8 March 2012). In this report, the GAO analyzes NASA's plan to take US$190 million from the Space Shuttle program to help offset additional costs of the ISS. The GAO notes that the transfer of the funds to the ISS program does not adversely affect current or near-term Shuttle upgrade projects.

U.S. General Accounting Office. "Space Station: Cost Control Difficulties Continue." Report no. GAO/NSIAD-96-135, Washington, DC, July 1996. *http://www.gao.gov/archive/1996/ ns96135.pdf* (accessed 8 March 2012). This report describes cost overruns on the ISS. As of April 1996, the prime contract for the ISS was nearly US$90 million over cost and about US$88 million behind schedule. If available resources prove inadequate, program managers either will be forced to exceed the annual funding limitation or will have to defer or rephase other activities, potentially delaying the space station's schedule and increasing its overall cost.

U.S. General Accounting Office. "Space Station: Cost To Operate After Assembly Is Uncertain." Report no. GAO/NSIAD-99-177, Washington, DC, August 1999. *http://www.gao.gov/ archive/1999/ns99177.pdf* (accessed 8 March 2012). This GAO report analyzes the cost of operating a completed ISS. NASA said that Shuttle flights should be allocated to the overall cost of operating the space station using a marginal cost of US$84 million per flight rather than an average cost per flight of US$435 million. The GAO believes that the average cost per flight more accurately represents the resources NASA will spend to run the space station.

U.S. General Accounting Office. "Space Station: Estimated Total U.S. Funding Requirements." Report no. GAO/NSIAD-95-163, Washington, DC, June 1995. *http://archive.gao.gov/ t2pbat1/154552.pdf* (accessed 7 March 2012). This report, which focuses on spending for the ISS, questions whether the Space Shuttle program can support the space station's launch and assembly requirements.

U.S. General Accounting Office. "Space Station: Impact of the Expanded Russian Role on Funding Research." Report no. GAO/NSIAD-94-220, Washington, DC, 21 June 1994. *http://archive.gao.gov/t2pbat3/151975.pdf* (accessed 7 March 2012). This report focuses on cost savings attributable to increased Russian participation in the ISS.

U.S. General Accounting Office. "Space Station: Impact of the Grounding of the Shuttle Fleet." Report no. GAO-03-1107, Washington, DC, September 2003. *http://www.gao.gov/new. items/d031107.pdf* (accessed 7 March 2012). This report discusses the status of the ISS's assembly and research plans, cost implications for the program, and concerns about whether NASA will be able to continue assembling the space station in the wake of the loss of Space Shuttle *Columbia.* The GAO finds that, because of NASA's grounding of the Shuttle fleet subsequent to the *Columbia* disaster, the space station will cost more and take longer to complete, delaying key research objectives. The partners must now rely on the limited payload capacity of Russia's spacecraft, Soyuz and Progress, to rotate crew and provide logistics support, placing the ISS in survival mode.

U.S. General Accounting Office. "Space Station: Information on National Security Applications and Cost." Report no. GAO/NSIAD-93-208, Washington, DC, May 1993. *http://archive. gao.gov/t2pbat5/149216.pdf* (accessed 8 March 2012). The report describes the cost of Space Station Freedom, a NASA project to construct a permanently piloted Earth-orbiting space station in the 1980s. Although approved by President Ronald W. Reagan and announced in the 1984 State of the Union Address, the proposed Space Station Freedom was never constructed or completed as originally designed. After several cutbacks, the remnants of the project became part of the ISS.

U.S. General Accounting Office. "Space Station: Program Instability and Cost Growth Continue Pending Redesign." Report no. GAO/NSAID-93-187, Washington, DC, 18 May 1993. *http://archive.gao.gov/t2pbat6/149194.pdf* (accessed 7 March 2012). This report describes cost estimates and real costs for Space Station Freedom. All Space Shuttle flights from FY 1988 through FY 2000 will be devoted to assembling the space station. The proposed Space Station Freedom was never constructed or completed as originally designed, but the remnants of the project became part of the ISS.

U.S. General Accounting Office. "Update on the Impact of the Expanded Russian Role." Report no. GAO/NSAID-94-248, Washington, DC, 29 July 1994. *http://archive.gao.gov/ t2pbat2/152266.pdf* (accessed 6 March 2012). This report describes the economic benefits of Russia's participation in the ISS. The GAO found that Russian participation in the space station program offered no net savings that NASA could use to fund other program areas and to accelerate the station's assembly schedule. In fact, Russian participation could add US$400 million in program funding requirements because of lower than anticipated hardware contributions.

U.S. Government Accountability Office. "NASA: More Knowledge Needed To Determine Best Alternatives To Provide Space Station Logistics Support." Report no. GAO-05-488, Washington, DC, May 2005. *http://www.gao.gov/new.items/d05488.pdf* (accessed 7 March 2012). This report reviews NASA's assessment that returning the Space Shuttle to flight was the best option for providing support to the ISS. Combining information gathered from commercial industry and a better definition of space station and Shuttle requirements, NASA officials agree there is an opportunity to perform a more comprehensive assessment of alternatives, especially for logistics missions late this decade.

Van de Haar, Rudolf. "STS-126: Mission Doubles Up Crew Capacity." *Spaceflight*, February 2009. This article discusses STS-126, Space Shuttle *Endeavour*'s mission to deliver construction equipment to the ISS and to service the Solar Alpha Rotary Joints. The author describes crew members, daily activities, launch preparation, and payloads during the 15-day mission.

Van de Haar, Rudolf, Rudolf van Beest, and Clive Simpson. "STS-121: Back to Back for Discovery." *Spaceflight*, September 2006. This article discusses STS-121, Space Shuttle *Discovery*'s mission to deliver equipment and supplies to the ISS and to demonstrate techniques for inspecting and protecting the Shuttle's thermal protection system. The author describes crew members, daily activities, launch preparation, and payloads during the 12-day mission.

Link to Part 1 (1970–1991), Chapter 13—The Shuttle in International Perspective

CHAPTER 18—THE END OF THE SPACE SHUTTLE PROGRAM

Asker, James R. "Radical Upgrades Urged To Cut Shuttle Costs." *Aviation Week & Space Technology*, 29 November 1993. This special report explains what NASA could do to keep the Space Shuttle flying until 2030. One idea is to replace the solid rocket boosters with winged, liquid-fuel boosters that would be flown back to land on a runway after they drop off the Shuttle orbiter's external tank stack two minutes into flight. The author also reviews possible changes to the Shuttle's exterior, reaction-control systems, auxiliary power units, and payload bay.

Brandon-Cremer, Lee, and Joel Powell. *Space Shuttle Almanac*. Calgary, AB: Microgravity Productions, 1992. *http://www.amazon.com/Space-Shuttle-Almanac-ebook/dp/ B005IWAVOA/ref=sr_1_8?s=books&ie=UTF8&qid=1323720909&sr=1-8* (accessed 8 March 2012). In this final digital edition of the *Space Shuttle Almanac*, primary author Lee Brandon-Cremer celebrates 40 years of Shuttle operational history within a 1,400-page compilation of mission facts, figures, dates, and times. The almanac includes an outstanding collection of more than 1,000 photographs and more than 1,000 diagrams, covering every mission. The e-book is available on CD or by download.

Broad, William J. "Fears of Decline and Demoralization at NASA as the Shuttle Program Ends." *New York Times*, 4 July 2011. As NASA prepares to launch its last Space Shuttle— ending 30 years in which large teams of creative scientists and engineers sent winged spacecraft into orbit—it faces a potentially greater challenge: a brain drain that threatens to undermine safety, as well as jeopardizing NASA's plans.

Brumfiel, Geoff. "Replacing the Space Shuttle: On Wings and a Prayer." *Nature* 421, no. 6924 (13 February 2003): 684. This article focuses on NASA's two-pronged effort to replace the Space Shuttles. One NASA effort would develop an Orbital Space Plane within 10 years, using conventional launch technologies. The other effort is an open-ended concept study with no time limit. Reducing fuel weight is a goal set for each effort.

Brumfiel, Geoff. "Space Exploration: Where 24 Men Have Gone Before." *Nature* 445, no. 7127 (2 January 2007): 474–478. The article discusses challenges to NASA's space exploration efforts and the new Space Shuttle craft that NASA plans to use by 2020.

Chang, Kenneth. "With 'Coolest Job Ever' Ending, Astronauts Seek Next Frontier." *New York Times*, 24 April 2011. The author reports that, as the Space Shuttle program ends, the likelihood of human spaceflight to an interesting destination anytime soon is decreasing. Consequently, astronauts are experiencing low morale.

Dicht, Burton. "Shuttle Diplomacy." *Mechanical Engineering*, July 2011. With the Space Shuttle era coming to a close and the final flight of *Atlantis* scheduled for July 2011, the author of this article discusses the history of the Space Shuttle program and the future of the U.S. human spaceflight program after it ends. Assessing the costs of the Space Shuttle and its contributions to the space program, the author suggests that, although the reusable spacecraft is a remarkable flying machine, it has not achieved the goal that led to its creation—to reduce the cost of delivering humans and large payloads into space.

Duggins, Pat. *Final Countdown: NASA and the End of the Space Shuttle Program*. Gainesville, FL: University Press of Florida, 2009. This book describes the Space Shuttle program from beginning to end, including the early Shuttle plans and many of the Space Shuttle missions. The author interviews Shuttle astronauts and writes about the impending demise of the Space Shuttle. The book also examines the plans for NASA's next major spacecraft system and the early phase of its development.

Fountain, Henry. "Shuttles, Turning Sedentary, Leave Pieces Behind for Science and Safety." *New York Times*, 2 June 2011. Although it is preparing to end the Space Shuttle program, NASA does not plan to ship the retired Shuttles to museums complete with all of their parts. Crews preparing for the retirement of the spacecraft are inundated with requests to retain many Shuttle parts for analysis, including the Shuttles' valves, flight-control instruments, and even their tires and windows.

Gale, Morrison. "Shuttle Diplomacy." *Mechanical Engineering*, March 1999. This article discusses how a National Research Council (NRC) report to NASA significantly influenced the policy decision that determined the fate of the Space Shuttle program. A special committee of the NRC recommended that NASA make 25 upgrades to the Space Shuttle, causing NASA to reconsider whether the Shuttle would continue to operate after 2012.

Griffin, Gerry. "As Shuttle Retires, a Vote for Commercial Space Flight." *USA Today*, 6 April 2011. The author reports that, because it will not have low-Earth-orbit transportation capability after the Space Shuttle program ends, NASA will not be able to explore and learn more about space. Furthermore, NASA will not be able to conduct piloted spaceflight once the Space Shuttle retires. However, the author believes that the commercial spaceflight industry shows encouraging signs that it may develop the capability of conducting human spaceflight in the near future.

Harpole, Tom. "Throttle Down." *Air & Space*, November 2010. This article discusses the prospective end of the Space Shuttle program and the closing of NASA's Kennedy Space Center. The author notes that closing Kennedy Space Center will lead to the loss of an estimated 9,000 NASA jobs and will have an effect on the economy of the region in Florida where it is located.

"Keep to the Vision." *Nature* 459, no. 7243 (July 5, 2009): 9–10. The article reflects on the possibility of continuing the Space Shuttle fleet past the year 2010.

Kremer, Ken. "Space Shuttle Processing Facilities." *Spaceflight*, April 2009. This article describes NASA's typical work process to ready a Shuttle for flight and to unload, inspect, repair, clean, and test the Shuttle upon its return to Earth. In addition, the author describes the facilities and equipment that NASA uses to process Shuttles, as well as the tasks performed at NASA's Kennedy Space Center between Space Shuttle launches. Once the Space Shuttle retires, NASA will no longer need staff to perform these tasks.

Lawler, Andrew. "After Columbia, a New NASA?" *Science* 299, no. 5609 (14 February 2003): 998. The author believes that the *Columbia* tragedy will result in a dialogue between Congress and NASA that will determine the direction of the U.S. space program over the next 20 years and will ultimately have a beneficial effect on space policy. NASA Administrator Sean O'Keefe hopes to build a complement to the Space Shuttle, to revolutionize space science missions, and to create the framework for human missions beyond Earth orbit.

Leary, Warren E. "NASA Starts Planning To Retire Space Shuttle." *New York Times*, 2 April 2005. Even as NASA prepared to resume flights of the Space Shuttle, a top NASA official reported that NASA had begun to form detailed plans to retire the spacecraft in five years, if not sooner.

Leary, Warren E. "Shuttle Retirement May Bring Loss of 8,600 Jobs, NASA Says." *New York Times*, 2 April 2008. In response to a congressional order, NASA released its first estimates of anticipated job losses during its transition from the Space Shuttle program to the Constellation program. NASA estimates the loss of 8,000 contractor jobs and 600 civil service jobs.

Martin, Benjamin P. "Space Shuttle Re-envisioned." *Ad Astra*, Spring 2007. This article focuses on the capabilities of a proposed new, unnamed orbiter like the Space Shuttle. According to the author, the new orbital spacecraft's fixed-wing concept does not use propellants. It will take off vertically, along with an attached tanker that will supply all of the propellants for both vehicles. The design of the new spacecraft diminishes the dynamic penalties of structures pushing against each other during their ascent trajectory.

Morgan, Daniel. "The Future of NASA: Space Policy Issues Facing Congress." CRS Report for Congress, Congressional Research Service, Library of Congress, Washington, DC, 7 August 2010. *http://assets.opencrs.com/rpts/R41016_20100708.pdf* (accessed 8 March 2012). The report reviews President George W. Bush's Vision for Space Exploration; the findings of the U.S. Human Space Flight Plans Committee, which President Bush appointed to report on the future of human spaceflight in the United States; President Barack H. Obama's cancellation of the Constellation program; and congressional hearings regarding the extension of the Space Shuttle program beyond 2011.

Morgan, Daniel, and Carl E. Behrens. "National Aeronautics and Space Administration: Overview, FY 2008 Budget in Brief, and Key Issues for Congress." CRS Report for Congress, Congressional Research Service, Library of Congress, Washington, DC, 14 March 2007. *http://assets.opencrs.com/rpts/RS22625_20070314.pdf* (accessed 8 March 2012). This report discusses the FY 2008 US$17.309 billion budget request for NASA, an increase of 6.5 percent from the FY 2007 appropriation of US$16.247 billion. Other issues addressed include the President's Vision for Space Exploration, development of new vehicles for human spaceflight, plans for the transition to these vehicles after NASA retires the Space Shuttle in 2010, and NASA's efforts to balance its priorities between human exploration and its other activities in science and aeronautics.

Morgan, Daniel, and Carl E. Behrens. "National Aeronautics and Space Administration: Overview, FY 2009 Budget, and Issues for Congress." CRS Report for Congress, Congressional Research Service, Library of Congress, Washington, DC, 26 February 2008. *http://assets.opencrs.com/rpts/RS22818_20080226.pdf* (accessed 8 March 2012). This report discusses NASA's FY 2009 budget request of US$17.614 billion, an increase of 1.8 percent from the FY 2008 appropriation of US$17.309 billion. The report explains the importance of implementing the President's Vision for Space Exploration, including the development of new vehicles for human spaceflight, plans for the transition to these vehicles after NASA retires the Space Shuttle in 2010, and NASA's efforts to balance its priorities between human exploration and its other activities in science and aeronautics.

Morring, Frank, Jr. "Commercial Break; NASA Plans COTS-Only Approach for ISS, Dropping Russia's Progress." *Aviation Week & Space Technology*, 21 April 2008. This article reports that NASA officials will discuss with Congress a plan for NASA to continue using Russia's Soyuz crew launch vehicles to transport astronauts to and from the ISS after the final Space Shuttle flight in 2010. NASA does not intend to continue using Russian Progress vehicles for U.S. cargo resupply but plans to use its own Commercial Orbital Transportation System (COTS) program vehicles, which are as yet untested.

Neal, Valerie. "Space Policy and the Size of the Space Shuttle Fleet." *Space Policy* 20, no. 3 (August 2004): 157–169. The author describes arguments for and against changes to the size of the Shuttle fleet and the influence of those arguments on NASA's policy and plans for after the retirement of the Space Shuttle. The author also discusses the size of the Shuttle fleet after the *Challenger* accident.

Reuters. "Panel Agrees To Protect Jobs, Add Mission for Space Shuttle." *Washington Post*, 16 July 2010. Reuters reports that the Senate Commerce, Science, and Transportation Committee has unanimously passed a plan to postpone retirement of the Space Shuttle. In addition, the committee agreed to include in NASA's three-year spending plan an additional Space Shuttle mission to the ISS.

Saslow, Rachel. "A Man Who Spent Years at NASA and 35 Days in Space." *Washington Post*, 5 July 2011. The author interviews former Space Shuttle astronaut Piers J. Sellers about his memories of spaceflight and the end of the U.S. Space Shuttle program.

Smith, Marcia S. "Space Exploration: Issues Concerning the 'Vision for Space Exploration.'" CRS Report for Congress, Congressional Research Service, Library of Congress, Washington, DC, 6 September 2005. *http://assets.opencrs.com/rpts/RS21720_20050609.pdf* (accessed 8 March 2012). This report provides an overview of President George W. Bush's 2004 Vision for Space Exploration and congressional reaction to the Vision, which includes terminating the Space Shuttle program in 2010.

Smith, Marcia S., and Daniel Morgan. "The National Aeronautics and Space Administration's FY 2005 Budget Request: Description, Analysis, and Issues for Congress." CRS Report for Congress, Congressional Research Service, Library of Congress, Washington, DC, 12 October 2004. *http://assets.opencrs.com/rpts/RL32676_20041210.pdf* (accessed 8 March 2012). The report describes NASA's FY 2005 budget of US$16.070 billion, a 4.5 percent increase over NASA's FY 2004 appropriation of US$15.378 billion. According to President George W. Bush's Vision for Space Exploration, NASA will focus its activities on returning humans to the Moon by 2020 and someday sending them to Mars and to "worlds beyond."

Smith, Marcia S., and Daniel Morgan. "The National Aeronautics and Space Administration's FY 2006 Budget Request: Description, Analysis, and Issues for Congress." CRS Report for Congress, Congressional Research Service, Library of Congress, Washington, DC, 17 November 2005. *http://assets.opencrs.com/rpts/RL32988_20051117.pdf* (accessed 8 March 2012). This report describes NASA's FY 2006 budget and the congressional debate over NASA's future programs. NASA requested US$16.456 billion, 2.4 percent more than the US$16.070 billion Congress appropriated in FY 2005. NASA Administrator Michael D. Griffin is accelerating development of a crew exploration vehicle.

Trabucco, Peter. "What's Next for NASA After the Space Shuttle?" *Ad Astra*, Fall 2010. This article focuses on the state of the U.S. space program and the future of NASA. Because of space program budget cuts, employees of NASA's Johnson Space Center could lose their jobs after the last Space Shuttle flight returns from space.

U.S. Congress. Congressional Budget Office. *A Budgetary Analysis of NASA's New Vision for Space Exploration*. Report, Washington, DC, 2 September 2004. *http://www.cbo.gov/ftpdocs/57xx/doc5772/09-02-NASA.pdf* (accessed 8 March 2012). In this report analyzing NASA's budget request for FY 2005 and NASA's budget projection through 2020, the CBO assesses the implications of NASA's budget plans on the content and schedule of NASA's future activities, including the operation of the Space Shuttle and the United States' participation in the ISS. Funding would enable NASA to develop new vehicles for spaceflight, allowing humans to return to the Moon by 2020.

U.S. Congress. Congressional Budget Office. *Analysis of NASA's Plans for Continuing Human Spaceflight After Retiring the Space Shuttle*. Report, Washington, DC, November 2008. *http://www.cbo.gov/ftpdocs/98xx/doc9886/11-03-NASA_Letter.pdf* (accessed 8 March 2012). This report describes the status of NASA's Constellation program, which funds the development of new vehicles, including the Ares-1 crew launch vehicle and the *Orion* crew exploration vehicle intended for future human spaceflight to the Moon, Mars, and beyond. The report also evaluates NASA's prospects for achieving its goals.

U.S. Congress. Congressional Budget Office. *The Budgetary Implications of NASA's Current Plans for Space Exploration*. Report, Washington, DC, April 2009. *http://www.cbo.gov/ftpdocs/100xx/doc10051/04-15-NASA.pdf* (accessed 8 March 2012). This report provides cost projections based on NASA's current plans for retiring the Space Shuttle and the extension of the Space Shuttle program into 2015. The extension would enable NASA to eliminate the gap between the Shuttle's retirement and the initial operating capability of other spacecraft, so that the United States' ability to conduct human spaceflight would not be compromised.

U.S. Congress. House of Representatives. Committee on Science. *Implementing the Vision for Space Exploration: Development of the Crew Exploration Vehicle.* 109th Cong., 2nd sess., 28 September 2006. *http://www.gpo.gov/fdsys/pkg/CHRG-109hhrg29949/pdf/CHRG-109hhrg29949.pdf* (accessed 12 March 2012). This hearing summarizes the development of the program that will build the *Orion* crew exploration vehicle, including an overview of potential challenges to the program's implementation and sustainability and a discussion of budgetary concerns. For the Orion program, which would rely on Shuttle-derived technologies, NASA must use Space Shuttle personnel, assets, and infrastructure to develop a crew exploration vehicle, a crew launch vehicle, and a heavy-lift launch vehicle.

U.S. Congress. House of Representatives. Committee on Science. Subcommittee on Space and Aeronautics. *Space Shuttle and Space Launch Initiative.* 107th Cong., 2nd sess., 18 April 2002. *https://web.lexis-nexis.com/congcomp/* (accessed 11 November 2011 via Proquest Congressional). This hearing reviews proposed Space Shuttle safety and performance upgrades and examines the NASA Space Launch Initiative, a program for research into the development and commercial applications of advanced and alternative space transportation technologies, including RLV development.

U.S. Congress. House of Representatives. Committee on Science. Subcommittee on Space and Aeronautics. *Space Transportation, Parts I–IV.* 106th Cong., 1st sess., 29 September; 13–27 October 1999. *https://web.lexis-nexis.com/congcomp/* (accessed 11 November 2011 via Proquest Congressional). This hearing reviews the status of NASA's RLV program, including the X-33 RLV demonstration program; assesses private-sector efforts to develop RLVs using private capital; evaluates proposed safety and performance upgrades to the Space Shuttle; and examines the development of future space transportation systems.

U.S. Congress. House of Representatives. Committee on Science and Technology. *Options and Issues for NASA's Human Space Flight Program: Report of the "Review of U.S. Human Space Flight Plans" Committee.* 111th Cong. 1st sess., 15 September 2009. *http://www.gpo.gov/fdsys/pkg/CHRG-111hhrg51928/pdf/CHRG-111hhrg51928.pdf* (accessed 6 March 2012). This hearing examines the findings of the U.S. Human Spaceflight Plans Committee's independent review of current U.S. human spaceflight plans and possible alternatives, identifying options to enable continued human spaceflight beyond the Space Shuttle's retirement.

U.S. Congress. House of Representatives. Committee on Science and Technology. Subcommittee on Space and Aeronautics. *NASA's International Space Station Program: Status and Issues.* 110th Cong., 2nd sess., 24 April 2008. *http://www.gpo.gov/fdsys/pkg/CHRG-110hhrg41799/pdf/CHRG-110hhrg41799.pdf* (accessed 7 March 2012). This hearing examines the challenges and risks facing the ISS program in light of the decision to retire the Space Shuttle in 2010.

U.S. Congress. Senate. Committee on Commerce, Science, and Transportation. Subcommittee on Science and Space. *Assessing Commercial Space Capabilities.* 111th Cong., 2nd sess., 18 March 2010. *http://frwebgate.access.gpo.gov/cgi-bin/getdoc.cgi?dbname=111_senate_ hearings&docid=f:66983.pdf* (accessed 14 March 2012). This hearing examines commercial space capabilities and developments in light of the President George W. Bush administration's decision to discontinue the Space Shuttle program in favor of expanded commercial-sector contracting, under NASA oversight, for transportation of astronauts to the ISS and for further exploration.

U.S. Congress. Senate. Committee on Commerce, Science, and Transportation. Subcommittee on Science and Space. *Options from the Review of U.S. Human Spaceflight Plans Committee.* 111th Cong., 1st sess., 16 September 2009. *http://frwebgate.access.gpo.gov/ cgi-bin/getdoc.cgi?dbname=111_senate_hearings&docid=f:54289.pdf* (accessed 12 March 2012). This hearing examines the findings of the U.S. Human Spaceflight Plans Committee's independent review of current U.S. human spaceflight plans and possible alternatives, identifying options to enable continued human spaceflight beyond the Space Shuttle's retirement.

U.S. Congress. Senate. Committee on Commerce, Science, and Transportation. Subcommittee on Space, Aeronautics, and Related Sciences. *Issues Facing the U.S. Space Program After Retirement of the Space Shuttle.* 110th Cong., 1st sess., 15 November 2007. In this hearing, NASA witnesses discuss various aspects of the U.S. space program after NASA retires the Space Shuttle, including the status of space transportation in support of the ISS.

U.S. Congress. Senate. Committee on Commerce, Science, and Transportation. Subcommittee on Space, Aeronautics, and Related Sciences. *Transitioning to a Next Generation Human Space Flight System.* 110th Cong., 1st sess., 28 March 2007. *http://www.gpo.gov/fdsys/ pkg/CHRG-110shrg39519/pdf/CHRG-110shrg39519.pdf* (accessed 14 March 2012). This hearing reviews NASA's plans for the transition from the Space Shuttle program to the Constellation program, including projects to develop the Ares crew launch vehicle and the *Orion* crew exploration vehicle. The hearing also examines NASA's efforts to minimize the anticipated time gap between the last Space Shuttle flight and the first planned launch using the new system. In addition, it assesses NASA's efforts to sustain the Space Shuttle workforce while developing the new spaceflight system.

U.S. General Accounting Office. "Space Transportation: Status of the X-33 Reusable Launch Vehicle Program." Report no. GAO/T-NSIAD-99-243, Washington, DC, 29 September 1999. *http://www.gao.gov/archive/1999/ns99243t.pdf* (accessed 8 March 2012). The report focuses on the possible phaseout of the Space Shuttle and its replacement with commercial launch services. The GAO remarks that the program will not meet some of its original cost, schedule, and performance objectives because of problems developing technologies for the X-33 VentureStar. In addition, the report states that the X-33 will not carry as much cargo as the Space Shuttle and will have to dock at the ISS more frequently than the Shuttle does.

U.S. Government Accountability Office. "NASA: Agency Has Taken Steps Toward Making Sound Investment Decisions for Ares I But Still Faces Challenging Knowledge Gaps." Report no. GAO-08-51, Washington, DC, October 2007. *http://www.gao.gov/new.items/ d0851.pdf* (accessed 8 March 2012). This report examines the Ares 1 and the *Orion* crew exploration vehicle as the replacement system for the Space Shuttle. NASA has not yet developed the knowledge necessary to make sound investment decisions for the Ares-1 project. The principal gaps in NASA's knowledge concern the project's requirements, costs, schedule, technology, design, and production feasibility. Continued instability in the design of the *Orion* crew exploration vehicle is hampering NASA's efforts to establish firm requirements for the Ares-1 project.

U.S. Government Accountability Office. "NASA: Progress Made on Strategic Human Capital Management, but Future Program Challenges Remain." Report no. GAO-07-1004, Washington, DC, 8 August 2007. *http://www.gao.gov/new.items/d071004.pdf* (accessed 6 March 2012). This report examines NASA's efforts to recruit, develop, and retain certain critical skills in its workforce, guided by its strategic human capital management plan, while working to replace the Space Shuttle with the next generation of human spaceflight systems. NASA is considering how best to mitigate the potential loss of skills and knowledge during the period from the Space Shuttle's retirement in 2010 to the resumption of human spaceflight in 2015.

U.S. Government Accountability Office. "NASA Supplier Base: Challenges Exist in Transitioning from the Space Shuttle Program to the Next Generation of Human Space Flight Systems." Report no. GAO-07-940, Washington, DC, July 2007. *http://www.gao. gov/new.items/d07940.pdf* (accessed 8 March 2012). This report examines NASA's plans and processes for managing its supplier base through the Shuttle's retirement and the transition to the Constellation program's exploration activities. The report recommends that the NASA Administrator instruct the Exploration Systems Mission Directorate and the Space Operations Mission Directorate to develop cost estimates for transition and retirement activities beyond FY 2010 jointly, so that NASA can include in its FY 2009 budget submission the transition and retirement funding through FY 2013.

U.S. Government Accountability Office. "Space Shuttle: Actions Needed To Better Position NASA To Sustain Its Workforce Through Retirement." Report no. GAO-05-230, Washington, DC, March 2005. *http://www.gao.gov/new.items/d05230.pdf* (accessed 6 March 2012). This GAO report discusses President George W. Bush's Vision for Space Exploration, which directs NASA to retire the Space Shuttle following completion of the ISS. The retirement process, which will last several years, will affect thousands of critically skilled NASA civil service employees and contractors supporting the Space Shuttle program. The key to implementing the Vision is NASA's ability to sustain its workforce to support safe Shuttle operations until NASA has completed the spacecraft's retirement.

Watson, Traci. "U.S. Human Spaceflight and the Road Ahead." *USA Today*, 26 January 2011. This short article describes Shuttle flights, from the beginning of the Shuttle era through the life of the Space Shuttle program, as well as outlining the future of the Shuttle and speculating on what will happen after NASA retires the spacecraft.

CHAPTER 19—MEMOIRS ABOUT THE SPACE SHUTTLE

Glenn, John H., Jr. *John Glenn: A Memoir*. New York: Bantam Books, 1999. This autobiography chronicles the life of astronaut John H. Glenn Jr., who is a veteran of World War II and the Korean War, as well as a former U.S. Senator. Glenn flew in NASA's Mercury program and, in 1962, became the first American to orbit the Earth. As a senior citizen, Glenn returned to space, flying on STS-95, a mission that focused on life sciences experiments.

Harris, Bernard A., Jr. *Dream Walker: A Journey of Achievement and Inspiration*. Austin, TX: Greenleaf Book Press Group, 2010. This book by Bernard A. Harris Jr., Mission Specialist on STS-55 and Payload Commander on STS-63, describes Harris's modest background and his experiences in college, medical school, and during training as a NASA flight surgeon. Harris was the first African American to walk in space.

Jones, Thomas D. *Sky Walking: An Astronaut's Memoir*. New York: HarperCollins, 2007. The author of this book, astronaut Thomas D. Jones, flew on several Space Shuttle missions— STS-59 and STS-68 in 1994, STS-80 in 1996, and STS-98 in 2001. During those missions, Jones spent more than 19 hours spacewalking outside the Space Shuttle. Jones writes of the excitement of launch and of the experience of working in space with cosmonauts from the former Soviet Union. The book also describes the difficulties of participating in astronaut training and the emotional toll of training and missions on astronauts' families.

Mullane, R. Mike. *Liftoff! An Astronaut's Dream*. Parsippany, NJ: Silver Burdett, 1995. The author, a former astronaut, describes his experiences in space and shares his ideas about the future of spaceflight.

Mullane, Mike. *Riding Rockets: The Outrageous Tales of a Space Shuttle Astronaut*. New York: Simon and Schuster, 2007. The author, former U.S. Air Force Colonel Richard Michael "Mike" Mullane, was Space Shuttle mission specialist on STS-27, STS-36, and STS-41. In his autobiography, Mullane describes clashes between astronauts with military experience and those with an academic background. He also discusses how his relationships with female astronauts changed over time.

Reichhardt, Tony, ed. *Space Shuttle: The First 20 Years—The Astronauts' Experiences in Their Own Words*. New York: Dorling Kindersley, 2002. This book, compiled by the editors of *Air & Space* and *Smithsonian* magazines, documents the history of the Space Shuttle program based on astronauts' anecdotes and reminiscences. The book includes 77 first-person accounts, including astronauts' descriptions of their experiences in zero gravity and their fear of failing in their missions.

Scott, Winston E. *Reflections from Earth Orbit*. Burlington, ON: Apogee, 2005. In this book, former naval aviator and astronaut Winston E. Scott recounts the obstacles he overcame to enter college, become a naval aviator, and gain entrance into astronaut training. Scott, who flew aboard STS-72 in 1996 and STS-87 in 1997 as a mission specialist describes the realities of living in space, emphasizing the routine aspects of life in space.

Link to Part 1 (1970–1991), Chapter 12—Shuttle Astronauts

THE NASA HISTORY SERIES

Reference Works, NASA SP-4000:

Grimwood, James M. *Project Mercury: A Chronology*. NASA SP-4001, 1963.

Grimwood, James M., and Barton C. Hacker, with Peter J. Vorzimmer. *Project Gemini Technology and Operations: A Chronology*. NASA SP-4002, 1969.

Link, Mae Mills. *Space Medicine in Project Mercury*. NASA SP-4003, 1965.

Astronautics and Aeronautics, 1963: Chronology of Science, Technology, and Policy. NASA SP-4004, 1964.

Astronautics and Aeronautics, 1964: Chronology of Science, Technology, and Policy. NASA SP-4005, 1965.

Astronautics and Aeronautics, 1965: Chronology of Science, Technology, and Policy. NASA SP-4006, 1966.

Astronautics and Aeronautics, 1966: Chronology of Science, Technology, and Policy. NASA SP-4007, 1967.

Astronautics and Aeronautics, 1967: Chronology of Science, Technology, and Policy. NASA SP-4008, 1968.

Ertel, Ivan D., and Mary Louise Morse. *The Apollo Spacecraft: A Chronology, Volume I, Through November 7, 1962*. NASA SP-4009, 1969.

Morse, Mary Louise, and Jean Kernahan Bays. *The Apollo Spacecraft: A Chronology, Volume II, November 8, 1962–September 30, 1964*. NASA SP-4009, 1973.

Brooks, Courtney G., and Ivan D. Ertel. *The Apollo Spacecraft: A Chronology, Volume III, October 1, 1964–January 20, 1966*. NASA SP-4009, 1973.

Ertel, Ivan D., and Roland W. Newkirk, with Courtney G. Brooks. *The Apollo Spacecraft: A Chronology, Volume IV, January 21, 1966–July 13, 1974*. NASA SP-4009, 1978.

Astronautics and Aeronautics, 1968: Chronology of Science, Technology, and Policy. NASA SP-4010, 1969.

Newkirk, Roland W., and Ivan D. Ertel, with Courtney G. Brooks. *Skylab: A Chronology*. NASA SP-4011, 1977.

Van Nimmen, Jane, and Leonard C. Bruno, with Robert L. Rosholt. *NASA Historical Data Book, Volume I: NASA Resources, 1958–1968*. NASA SP-4012, 1976; rep. ed. 1988.

Ezell, Linda Neuman. *NASA Historical Data Book, Volume II: Programs and Projects, 1958–1968.* NASA SP-4012, 1988.

Ezell, Linda Neuman. *NASA Historical Data Book, Volume III: Programs and Projects, 1969–1978.* NASA SP-4012, 1988.

Gawdiak, Ihor, with Helen Fedor. *NASA Historical Data Book, Volume IV: NASA Resources, 1969–1978.* NASA SP-4012, 1994.

Rumerman, Judy A. *NASA Historical Data Book, Volume V: NASA Launch Systems, Space Transportation, Human Spaceflight, and Space Science, 1979–1988.* NASA SP-4012, 1999.

Rumerman, Judy A. *NASA Historical Data Book, Volume VI: NASA Space Applications, Aeronautics and Space Research and Technology, Tracking and Data Acquisition/Support Operations, Commercial Programs, and Resources, 1979–1988.* NASA SP-4012, 1999.

Rumerman, Judy A. *NASA Historical Data Book, Volume VII: NASA Launch Systems, Space Transportation, Human Spaceflight, and Space Science, 1989–1998.* NASA SP-2009-4012, 2009.

Rumerman, Judy A. *NASA Historical Data Book, Volume VIII: NASA Earth Science and Space Applications, Aeronautics, Technology, and Exploration, Tracking and Data Acquisition/Space Operations, Facilities and Resources, 1989–1998.* NASA SP-2012-4012, 2012.

No SP-4013.

Astronautics and Aeronautics, 1969: Chronology of Science, Technology, and Policy. NASA SP-4014, 1970.

Astronautics and Aeronautics, 1970: Chronology of Science, Technology, and Policy. NASA SP-4015, 1972.

Astronautics and Aeronautics, 1971: Chronology of Science, Technology, and Policy. NASA SP-4016, 1972.

Astronautics and Aeronautics, 1972: Chronology of Science, Technology, and Policy. NASA SP-4017, 1974.

Astronautics and Aeronautics, 1973: Chronology of Science, Technology, and Policy. NASA SP-4018, 1975.

Astronautics and Aeronautics, 1974: Chronology of Science, Technology, and Policy. NASA SP-4019, 1977.

Astronautics and Aeronautics, 1975: Chronology of Science, Technology, and Policy. NASA SP-4020, 1979.

Astronautics and Aeronautics, 1976: Chronology of Science, Technology, and Policy. NASA SP-4021, 1984.

Astronautics and Aeronautics, 1977: Chronology of Science, Technology, and Policy. NASA SP-4022, 1986.

Astronautics and Aeronautics, 1978: Chronology of Science, Technology, and Policy. NASA SP-4023, 1986.

Astronautics and Aeronautics, 1979–1984: Chronology of Science, Technology, and Policy. NASA SP-4024, 1988.

Astronautics and Aeronautics, 1985: Chronology of Science, Technology, and Policy. NASA SP-4025, 1990.

Noordung, Hermann. *The Problem of Space Travel: The Rocket Motor.* Edited by Ernst Stuhlinger and J. D. Hunley, with Jennifer Garland. NASA SP-4026, 1995.

Gawdiak, Ihor Y., Ramon J. Miro, and Sam Stueland. *Astronautics and Aeronautics, 1986–1990: A Chronology.* NASA SP-4027, 1997.

Gawdiak, Ihor Y., and Charles Shetland. *Astronautics and Aeronautics, 1991–1995: A Chronology.* NASA SP-2000-4028, 2000.

Orloff, Richard W. *Apollo by the Numbers: A Statistical Reference.* NASA SP-2000-4029, 2000.

Lewis, Marieke, and Ryan Swanson. *Astronautics and Aeronautics: A Chronology, 1996–2000.* NASA SP-2009-4030, 2009.

Ivey, William Noel, and Marieke Lewis. *Astronautics and Aeronautics: A Chronology, 2001–2005.* NASA SP-2010-4031, 2010.

Buchalter, Alice R., and William Noel Ivey. *Astronautics and Aeronautics: A Chronology, 2006.* NASA SP-2011-4032, 2010.

Lewis, Marieke. *Astronautics and Aeronautics: A Chronology, 2007.* NASA SP-2011-4033, 2011.

Lewis, Marieke. *Astronautics and Aeronautics: A Chronology, 2008.* NASA SP-2012-4034, 2012.

Lewis, Marieke. *Astronautics and Aeronautics: A Chronology, 2009.* NASA SP-2012-4035, 2012.

Management Histories, NASA SP-4100:

Rosholt, Robert L. *An Administrative History of NASA, 1958–1963*. NASA SP-4101, 1966.

Levine, Arnold S. *Managing NASA in the Apollo Era*. NASA SP-4102, 1982.

Roland, Alex. *Model Research: The National Advisory Committee for Aeronautics, 1915–1958*. NASA SP-4103, 1985.

Fries, Sylvia D. *NASA Engineers and the Age of Apollo*. NASA SP-4104, 1992.

Glennan, T. Keith. *The Birth of NASA: The Diary of T. Keith Glennan*. Edited by J. D. Hunley. NASA SP-4105, 1993.

Seamans, Robert C. *Aiming at Targets: The Autobiography of Robert C. Seamans*. NASA SP-4106, 1996.

Garber, Stephen J., ed. *Looking Backward, Looking Forward: Forty Years of Human Spaceflight Symposium*. NASA SP-2002-4107, 2002.

Mallick, Donald L., with Peter W. Merlin. *The Smell of Kerosene: A Test Pilot's Odyssey*. NASA SP-4108, 2003.

Iliff, Kenneth W., and Curtis L. Peebles. *From Runway to Orbit: Reflections of a NASA Engineer*. NASA SP-2004-4109, 2004.

Chertok, Boris. *Rockets and People, Volume I*. NASA SP-2005-4110, 2005.

Chertok, Boris. *Rockets and People: Creating a Rocket Industry, Volume II*. NASA SP-2006-4110, 2006.

Chertok, Boris. *Rockets and People: Hot Days of the Cold War, Volume III*. NASA SP-2009-4110, 2009.

Chertok, Boris. *Rockets and People: The Moon Race, Volume IV*. NASA SP-2011-4110, 2011.

Laufer, Alexander, Todd Post, and Edward Hoffman. *Shared Voyage: Learning and Unlearning from Remarkable Projects*. NASA SP-2005-4111, 2005.

Dawson, Virginia P., and Mark D. Bowles. *Realizing the Dream of Flight: Biographical Essays in Honor of the Centennial of Flight, 1903–2003*. NASA SP-2005-4112, 2005.

Mudgway, Douglas J. *William H. Pickering: America's Deep Space Pioneer*. NASA SP-2008-4113, 2008.

Wright, Rebecca, Sandra Johnson, and Steven J. Dick. *NASA at 50: Interviews with NASA's Senior Leadership*. NASA SP-2012-4114, 2012.

Project Histories, NASA SP-4200:

Swenson, Loyd S., Jr., James M. Grimwood, and Charles C. Alexander. *This New Ocean: A History of Project Mercury*. NASA SP-4201, 1966; rep. ed. 1999.

Green, Constance McLaughlin, and Milton Lomask. *Vanguard: A History*. NASA SP-4202, 1970; rep. ed. Smithsonian Institution Press, 1971.

Hacker, Barton C., and James M. Grimwood. *On the Shoulders of Titans: A History of Project Gemini*. NASA SP-4203, 1977; rep. ed. 2002.

Benson, Charles D., and William Barnaby Faherty. *Moonport: A History of Apollo Launch Facilities and Operations*. NASA SP-4204, 1978.

Brooks, Courtney G., James M. Grimwood, and Loyd S. Swenson, Jr. *Chariots for Apollo: A History of Manned Lunar Spacecraft*. NASA SP-4205, 1979.

Bilstein, Roger E. *Stages to Saturn: A Technological History of the Apollo/Saturn Launch Vehicles*. NASA SP-4206, 1980 and 1996.

No SP-4207.

Compton, W. David, and Charles D. Benson. *Living and Working in Space: A History of Skylab*. NASA SP-4208, 1983.

Ezell, Edward Clinton, and Linda Neuman Ezell. *The Partnership: A History of the Apollo-Soyuz Test Project*. NASA SP-4209, 1978.

Hall, R. Cargill. *Lunar Impact: A History of Project Ranger*. NASA SP-4210, 1977.

Newell, Homer E. *Beyond the Atmosphere: Early Years of Space Science*. NASA SP-4211, 1980.

Ezell, Edward Clinton, and Linda Neuman Ezell. *On Mars: Exploration of the Red Planet, 1958–1978*. NASA SP-4212, 1984.

Pitts, John A. *The Human Factor: Biomedicine in the Manned Space Program to 1980*. NASA SP-4213, 1985.

Compton, W. David. *Where No Man Has Gone Before: A History of Apollo Lunar Exploration Missions*. NASA SP-4214, 1989.

Naugle, John E. *First Among Equals: The Selection of NASA Space Science Experiments*. NASA SP-4215, 1991.

Wallace, Lane E. *Airborne Trailblazer: Two Decades with NASA Langley's 737 Flying Laboratory*. NASA SP-4216, 1994.

Butrica, Andrew J., ed. *Beyond the Ionosphere: Fifty Years of Satellite Communications*. NASA SP-4217, 1997.

Butrica, Andrew J. *To See the Unseen: A History of Planetary Radar Astronomy.* NASA SP-4218, 1996.

Mack, Pamela E., ed. *From Engineering Science to Big Science: The NACA and NASA Collier Trophy Research Project Winners.* NASA SP-4219, 1998.

Reed, R. Dale. *Wingless Flight: The Lifting Body Story.* NASA SP-4220, 1998.

Heppenheimer, T. A. *The Space Shuttle Decision: NASA's Search for a Reusable Space Vehicle.* NASA SP-4221, 1999.

Hunley, J. D., ed. *Toward Mach 2: The Douglas D-558 Program.* NASA SP-4222, 1999.

Swanson, Glen E., ed. *"Before This Decade Is Out…" Personal Reflections on the Apollo Program.* NASA SP-4223, 1999.

Tomayko, James E. *Computers Take Flight: A History of NASA's Pioneering Digital Fly-By-Wire Project.* NASA SP-4224, 2000.

Morgan, Clay. *Shuttle-Mir: The United States and Russia Share History's Highest Stage.* NASA SP-2001-4225, 2001.

Leary, William M. *"We Freeze to Please": A History of NASA's Icing Research Tunnel and the Quest for Safety.* NASA SP-2002-4226, 2002.

Mudgway, Douglas J. *Uplink-Downlink: A History of the Deep Space Network, 1957–1997.* NASA SP-2001-4227, 2001.

No SP-4228 or SP-4229.

Dawson, Virginia P., and Mark D. Bowles. *Taming Liquid Hydrogen: The Centaur Upper Stage Rocket, 1958–2002.* NASA SP-2004-4230, 2004.

Meltzer, Michael. *Mission to Jupiter: A History of the Galileo Project.* NASA SP-2007-4231, 2007.

Heppenheimer, T. A. *Facing the Heat Barrier: A History of Hypersonics.* NASA SP-2007-4232, 2007.

Tsiao, Sunny. *"Read You Loud and Clear!" The Story of NASA's Spaceflight Tracking and Data Network.* NASA SP-2007-4233, 2007.

Meltzer, Michael. *When Biospheres Collide: A History of NASA's Planetary Protection Programs.* NASA SP-2011-4234, 2011.

Center Histories, NASA SP-4300:

Rosenthal, Alfred. *Venture into Space: Early Years of Goddard Space Flight Center.* NASA SP-4301, 1985.

Hartman, Edwin P. *Adventures in Research: A History of Ames Research Center, 1940–1965*. NASA SP-4302, 1970.

Hallion, Richard P. *On the Frontier: Flight Research at Dryden, 1946–1981*. NASA SP-4303, 1984.

Muenger, Elizabeth A. *Searching the Horizon: A History of Ames Research Center, 1940–1976*. NASA SP-4304, 1985.

Hansen, James R. *Engineer in Charge: A History of the Langley Aeronautical Laboratory, 1917–1958*. NASA SP-4305, 1987.

Dawson, Virginia P. *Engines and Innovation: Lewis Laboratory and American Propulsion Technology*. NASA SP-4306, 1991.

Dethloff, Henry C. *"Suddenly Tomorrow Came…": A History of the Johnson Space Center, 1957–1990*. NASA SP-4307, 1993.

Hansen, James R. *Spaceflight Revolution: NASA Langley Research Center from Sputnik to Apollo*. NASA SP-4308, 1995.

Wallace, Lane E. *Flights of Discovery: An Illustrated History of the Dryden Flight Research Center*. NASA SP-4309, 1996.

Herring, Mack R. *Way Station to Space: A History of the John C. Stennis Space Center*. NASA SP-4310, 1997.

Wallace, Harold D., Jr. *Wallops Station and the Creation of an American Space Program*. NASA SP-4311, 1997.

Wallace, Lane E. *Dreams, Hopes, Realities. NASA's Goddard Space Flight Center: The First Forty Years*. NASA SP-4312, 1999.

Dunar, Andrew J., and Stephen P. Waring. *Power to Explore: A History of Marshall Space Flight Center, 1960–1990*. NASA SP-4313, 1999.

Bugos, Glenn E. *Atmosphere of Freedom: Sixty Years at the NASA Ames Research Center*. NASA SP-2000-4314, 2000.

No SP-4315.

Schultz, James. *Crafting Flight: Aircraft Pioneers and the Contributions of the Men and Women of NASA Langley Research Center*. NASA SP-2003-4316, 2003.

Bowles, Mark D. *Science in Flux: NASA's Nuclear Program at Plum Brook Station, 1955–2005*. NASA SP-2006-4317, 2006.

Wallace, Lane E. *Flights of Discovery: An Illustrated History of the Dryden Flight Research Center*. NASA SP-2007-4318, 2007. Revised version of NASA SP-4309.

Arrighi, Robert S. *Revolutionary Atmosphere: The Story of the Altitude Wind Tunnel and the Space Power Chambers.* NASA SP-2010-4319, 2010.

Bugos, Glenn E. *Atmosphere of Freedom: Seventy Years at the NASA Ames Research Center.* NASA SP-2010-4314, 2010. Revised Version of NASA SP-2000-4314.

General Histories, NASA SP-4400:

Corliss, William R. *NASA Sounding Rockets, 1958–1968: A Historical Summary.* NASA SP-4401, 1971.

Wells, Helen T., Susan H. Whiteley, and Carrie Karegeannes. *Origins of NASA Names.* NASA SP-4402, 1976.

Anderson, Frank W., Jr. *Orders of Magnitude: A History of NACA and NASA, 1915–1980.* NASA SP-4403, 1981.

Sloop, John L. *Liquid Hydrogen as a Propulsion Fuel, 1945–1959.* NASA SP-4404, 1978.

Roland, Alex. *A Spacefaring People: Perspectives on Early Spaceflight.* NASA SP-4405, 1985.

Bilstein, Roger E. *Orders of Magnitude: A History of the NACA and NASA, 1915–1990.* NASA SP-4406, 1989.

Logsdon, John M., ed., with Linda J. Lear, Jannelle Warren Findley, Ray A. Williamson, and Dwayne A. Day. *Exploring the Unknown: Selected Documents in the History of the U.S. Civil Space Program, Volume I: Organizing for Exploration.* NASA SP-4407, 1995.

Logsdon, John M., ed., with Dwayne A. Day and Roger D. Launius. *Exploring the Unknown: Selected Documents in the History of the U.S. Civil Space Program, Volume II: External Relationships.* NASA SP-4407, 1996.

Logsdon, John M., ed., with Roger D. Launius, David H. Onkst, and Stephen J. Garber. *Exploring the Unknown: Selected Documents in the History of the U.S. Civil Space Program, Volume III: Using Space.* NASA SP-4407, 1998.

Logsdon, John M., ed., with Ray A. Williamson, Roger D. Launius, Russell J. Acker, Stephen J. Garber, and Jonathan L. Friedman. *Exploring the Unknown: Selected Documents in the History of the U.S. Civil Space Program, Volume IV: Accessing Space.* NASA SP-4407, 1999.

Logsdon, John M., ed., with Amy Paige Snyder, Roger D. Launius, Stephen J. Garber, and Regan Anne Newport. *Exploring the Unknown: Selected Documents in the History of the U.S. Civil Space Program, Volume V: Exploring the Cosmos.* NASA SP-2001-4407, 2001.

Logsdon, John M., ed., with Stephen J. Garber, Roger D. Launius, and Ray A. Williamson. *Exploring the Unknown: Selected Documents in the History of the U.S. Civil Space Program, Volume VI: Space and Earth Science*. NASA SP-2004-4407, 2004.

Logsdon, John M., ed., with Roger D. Launius. *Exploring the Unknown: Selected Documents in the History of the U.S. Civil Space Program, Volume VII: Human Spaceflight: Projects Mercury, Gemini, and Apollo*. NASA SP-2008-4407, 2008.

Siddiqi, Asif A., *Challenge to Apollo: The Soviet Union and the Space Race, 1945–1974*. NASA SP-2000-4408, 2000.

Hansen, James R., ed. *The Wind and Beyond: Journey into the History of Aerodynamics in America, Volume 1: The Ascent of the Airplane*. NASA SP-2003-4409, 2003.

Hansen, James R., ed. *The Wind and Beyond: Journey into the History of Aerodynamics in America, Volume 2: Reinventing the Airplane*. NASA SP-2007-4409, 2007.

Hogan, Thor. *Mars Wars: The Rise and Fall of the Space Exploration Initiative*. NASA SP-2007-4410, 2007.

Vakoch, Douglas A., ed. *Psychology of Space Exploration: Contemporary Research in Historical Perspective*. NASA SP-2011-4411, 2011.

Monographs in Aerospace History, NASA SP-4500:

Launius, Roger D., and Aaron K. Gillette, comps. *Toward a History of the Space Shuttle: An Annotated Bibliography*. Monographs in Aerospace History, No. 1, 1992.

Launius, Roger D., and J. D. Hunley, comps. *An Annotated Bibliography of the Apollo Program*. Monographs in Aerospace History, No. 2, 1994.

Launius, Roger D. *Apollo: A Retrospective Analysis*. Monographs in Aerospace History, No. 3, 1994.

Hansen, James R. *Enchanted Rendezvous: John C. Houbolt and the Genesis of the Lunar-Orbit Rendezvous Concept*. Monographs in Aerospace History, No. 4, 1995.

Gorn, Michael H. *Hugh L. Dryden's Career in Aviation and Space*. Monographs in Aerospace History, No. 5, 1996.

Powers, Sheryll Goecke. *Women in Flight Research at NASA Dryden Flight Research Center from 1946 to 1995*. Monographs in Aerospace History, No. 6, 1997.

Portree, David S. F., and Robert C. Trevino. *Walking to Olympus: An EVA Chronology*. Monographs in Aerospace History, No. 7, 1997.

Logsdon, John M., moderator. *Legislative Origins of the National Aeronautics and Space Act of 1958: Proceedings of an Oral History Workshop.* Monographs in Aerospace History, No. 8, 1998.

Rumerman, Judy A., comp. *U.S. Human Spaceflight: A Record of Achievement, 1961–1998.* Monographs in Aerospace History, No. 9, 1998.

Portree, David S. F. *NASA's Origins and the Dawn of the Space Age.* Monographs in Aerospace History, No. 10, 1998.

Logsdon, John M. *Together in Orbit: The Origins of International Cooperation in the Space Station.* Monographs in Aerospace History, No. 11, 1998.

Phillips, W. Hewitt. *Journey in Aeronautical Research: A Career at NASA Langley Research Center.* Monographs in Aerospace History, No. 12, 1998.

Braslow, Albert L. *A History of Suction-Type Laminar-Flow Control with Emphasis on Flight Research.* Monographs in Aerospace History, No. 13, 1999.

Logsdon, John M., moderator. *Managing the Moon Program: Lessons Learned from Apollo.* Monographs in Aerospace History, No. 14, 1999.

Perminov, V. G. *The Difficult Road to Mars: A Brief History of Mars Exploration in the Soviet Union.* Monographs in Aerospace History, No. 15, 1999.

Tucker, Tom. *Touchdown: The Development of Propulsion Controlled Aircraft at NASA Dryden.* Monographs in Aerospace History, No. 16, 1999.

Maisel, Martin, Demo J. Giulanetti, and Daniel C. Dugan. *The History of the XV-15 Tilt Rotor Research Aircraft: From Concept to Flight.* Monographs in Aerospace History, No. 17, 2000. NASA SP-2000-4517.

Jenkins, Dennis R. *Hypersonics Before the Shuttle: A Concise History of the X-15 Research Airplane.* Monographs in Aerospace History, No. 18, 2000. NASA SP-2000-4518.

Chambers, Joseph R. *Partners in Freedom: Contributions of the Langley Research Center to U.S. Military Aircraft of the 1990s.* Monographs in Aerospace History, No. 19, 2000. NASA SP-2000-4519.

Waltman, Gene L. *Black Magic and Gremlins: Analog Flight Simulations at NASA's Flight Research Center.* Monographs in Aerospace History, No. 20, 2000. NASA SP-2000-4520.

Portree, David S. F. *Humans to Mars: Fifty Years of Mission Planning, 1950–2000.* Monographs in Aerospace History, No. 21, 2001. NASA SP-2001-4521.

Thompson, Milton O., with J. D. Hunley. *Flight Research: Problems Encountered and What They Should Teach Us*. Monographs in Aerospace History, No. 22, 2001. NASA SP-2001-4522.

Tucker, Tom. *The Eclipse Project*. Monographs in Aerospace History, No. 23, 2001. NASA SP-2001-4523.

Siddiqi, Asif A. *Deep Space Chronicle: A Chronology of Deep Space and Planetary Probes, 1958–2000*. Monographs in Aerospace History, No. 24, 2002. NASA SP-2002-4524.

Merlin, Peter W. *Mach 3+: NASA/USAF YF-12 Flight Research, 1969–1979*. Monographs in Aerospace History, No. 25, 2001. NASA SP-2001-4525.

Anderson, Seth B. *Memoirs of an Aeronautical Engineer: Flight Tests at Ames Research Center: 1940–1970*. Monographs in Aerospace History, No. 26, 2002. NASA SP-2002-4526.

Renstrom, Arthur G. *Wilbur and Orville Wright: A Bibliography Commemorating the One-Hundredth Anniversary of the First Powered Flight on December 17, 1903*. Monographs in Aerospace History, No. 27, 2002. NASA SP-2002-4527.

No monograph 28.

Chambers, Joseph R. *Concept to Reality: Contributions of the NASA Langley Research Center to U.S. Civil Aircraft of the 1990s*. Monographs in Aerospace History, No. 29, 2003. NASA SP-2003-4529.

Peebles, Curtis, ed. *The Spoken Word: Recollections of Dryden History, The Early Years*. Monographs in Aerospace History, No. 30, 2003. NASA SP-2003-4530.

Jenkins, Dennis R., Tony Landis, and Jay Miller. *American X-Vehicles: An Inventory—X-1 to X-50*. Monographs in Aerospace History, No. 31, 2003. NASA SP-2003-4531.

Renstrom, Arthur G. *Wilbur and Orville Wright: A Chronology Commemorating the One-Hundredth Anniversary of the First Powered Flight on December 17, 1903*. Monographs in Aerospace History, No. 32, 2003. NASA SP-2003-4532.

Bowles, Mark D., and Robert S. Arrighi. *NASA's Nuclear Frontier: The Plum Brook Research Reactor*. Monographs in Aerospace History, No. 33, 2004. NASA SP-2004-4533.

Wallace, Lane, and Christian Gelzer. *Nose Up: High Angle-of-Attack and Thrust Vectoring Research at NASA Dryden, 1979–2001*. Monographs in Aerospace History, No. 34, 2009. NASA SP-2009-4534.

Matranga, Gene J., C. Wayne Ottinger, Calvin R. Jarvis, and D. Christian Gelzer. *Unconventional, Contrary, and Ugly: The Lunar Landing Research Vehicle*. Monographs in Aerospace History, No. 35, 2006. NASA SP-2004-4535.

McCurdy, Howard E. *Low-Cost Innovation in Spaceflight: The History of the Near Earth Asteroid Rendezvous (NEAR) Mission.* Monographs in Aerospace History, No. 36, 2005. NASA SP-2005-4536.

Seamans, Robert C., Jr. *Project Apollo: The Tough Decisions.* Monographs in Aerospace History, No. 37, 2005. NASA SP-2005-4537.

Lambright, W. Henry. *NASA and the Environment: The Case of Ozone Depletion.* Monographs in Aerospace History, No. 38, 2005. NASA SP-2005-4538.

Chambers, Joseph R. *Innovation in Flight: Research of the NASA Langley Research Center on Revolutionary Advanced Concepts for Aeronautics.* Monographs in Aerospace History, No. 39, 2005. NASA SP-2005-4539.

Phillips, W. Hewitt. *Journey into Space Research: Continuation of a Career at NASA Langley Research Center.* Monographs in Aerospace History, No. 40, 2005. NASA SP-2005-4540.

Rumerman, Judy A., Chris Gamble, and Gabriel Okolski, comps. *U.S. Human Spaceflight: A Record of Achievement, 1961–2006.* Monographs in Aerospace History, No. 41, 2007. NASA SP-2007-4541.

Peebles, Curtis. *The Spoken Word: Recollections of Dryden History Beyond the Sky.* Monographs in Aerospace History, No. 42, 2011. NASA SP-2011-4542.

Dick, Steven J., Stephen J. Garber, and Jane H. Odom. *Research in NASA History.* Monographs in Aerospace History, No. 43, 2009. NASA SP-2009-4543.

Merlin, Peter W. *Ikhana: Unmanned Aircraft System Western States Fire Missions.* Monographs in Aerospace History, No. 44, 2009. NASA SP-2009-4544.

Fisher, Steven C., and Shamim A. Rahman. *Remembering the Giants: Apollo Rocket Propulsion Development.* Monographs in Aerospace History, No. 45, 2009. NASA SP-2009-4545.

Gelzer, Christian. *Fairing Well: From Shoebox to Bat Truck and Beyond, Aerodynamic Truck Research at NASA's Dryden Flight Research Center.* Monographs in Aerospace History, No. 46, 2011. NASA SP-2011-4546.

Arrighi, Robert. *Pursuit of Power: NASA's Propulsion Systems Laboratory No. 1 and 2.* Monographs in Aerospace History, No. 48, 2012. NASA SP-2012-4548.

Goodrich, Malinda K., Alice R. Buchalter, and Patrick M. Miller, comps. *Toward a History of the Space Shuttle: An Annotated Bibliography, Part 2 (1992–2011).* Monographs in Aerospace History, No. 49, 2012. NASA SP-2012-4549.

Electronic Media, NASA SP-4600:

Remembering Apollo 11: The 30th Anniversary Data Archive CD-ROM. NASA SP-4601, 1999.

Remembering Apollo 11: The 35th Anniversary Data Archive CD-ROM. NASA SP-2004-4601, 2004. This is an update of the 1999 edition.

The Mission Transcript Collection: U.S. Human Spaceflight Missions from Mercury Redstone 3 to Apollo 17. NASA SP-2000-4602, 2001.

Shuttle-Mir: The United States and Russia Share History's Highest Stage. NASA SP-2001-4603, 2002.

U.S. Centennial of Flight Commission Presents Born of Dreams—Inspired by Freedom. NASA SP-2004-4604, 2004.

Of Ashes and Atoms: A Documentary on the NASA Plum Brook Reactor Facility. NASA SP-2005-4605, 2005.

Taming Liquid Hydrogen: The Centaur Upper Stage Rocket Interactive CD-ROM. NASA SP-2004-4606, 2004.

Fueling Space Exploration: The History of NASA's Rocket Engine Test Facility DVD. NASA SP-2005-4607, 2005.

Altitude Wind Tunnel at NASA Glenn Research Center: An Interactive History CD-ROM. NASA SP-2008-4608, 2008.

A Tunnel Through Time: The History of NASA's Altitude Wind Tunnel. NASA SP-2010-4609, 2010.

Conference Proceedings, NASA SP-4700:

Dick, Steven J., and Keith Cowing, eds. *Risk and Exploration: Earth, Sea and the Stars.* NASA SP-2005-4701, 2005.

Dick, Steven J., and Roger D. Launius. *Critical Issues in the History of Spaceflight.* NASA SP-2006-4702, 2006.

Dick, Steven J., ed. *Remembering the Space Age: Proceedings of the 50th Anniversary Conference.* NASA SP-2008-4703, 2008.

Dick, Steven J., ed. *NASA's First 50 Years: Historical Perspectives.* NASA SP-2010-4704, 2010.

Societal Impact, NASA SP-4800:

Dick, Steven J., and Roger D. Launius. *Societal Impact of Spaceflight.* NASA SP-2007-4801, 2007.

Dick, Steven J., and Mark L. Lupisella. *Cosmos and Culture: Cultural Evolution in a Cosmic Context.* NASA SP-2009-4802, 2009.

NASA SP-2012-4549

www.nasa.gov